The twentieth century has seen biology come of age as a conceptual and quantitative science. Biochemistry, cytology, and genetics have been unified into a common framework at the molecular level. However, cellular activity and development are regulated not by the interplay of molecules alone, but by interactions of molecules organized in complex arrays, subunits, and organelles. Emphasis on organization is, therefore, of increasing importance.

So it is too, at the other end of the scale. Organismic and population biology are developing new rigor in such established and emerging disciplines as ecology, evolution, and ethology, but again the accent is on interactions between individuals, populations, and societies. Advances in comparative biochemistry and physiology have given new impetus to studies of animal and plant diversity. Microbiology has matured, with the world of viruses and procaryotes assuming a major position. New connections are being forged with other disciplines outside biology—chemistry, physics, mathematics, geology, anthropology, and psychology provide us with new theories and experimental tools while at the same time are themselves being enriched by the biologists' new insights into the world of life. The need to preserve a habitable environment for future generations should encourage increasing collaboration between diverse disciplines.

The purpose of the Modern Biology Series is to introduce the college biology student—as well as the gifted secondary student and all interested readers—both to the concepts unifying the fields within biology and to the diversity that makes each field unique.

Since the series is open-ended, it will provide a greater number and variety of topics than can be accommodated in many introductory courses. It remains the responsibility of the instructor to make his selection, to arrange it in a logical order, and to develop a framework into which the individual units can best be fitted.

New titles will be added to the present list as new fields emerge, existing fields advance, and new authors of ability and talent appear. Only thus, we feel, can we keep pace with the explosion of knowledge in Modern Biology.

James D. Ebert
Ariel G. Loewy
Howard A. Schneiderman

Modern Biology Series

Consulting Editors

Published Titles

Burnett-Eisner	***Animal Adaptation***
Delevoryas	***Plant Diversification***
Ebert	***Interacting Systems in Development***
Loewy-Siekevitz	***Cell Structure and Function***
Odum	***Ecology***
Ray	***The Living Plant***
Van der Kloot	***Behavior***

Forthcoming Titles

Fingerman	***Animal Diversity***
Gillespie	***Cell Function***
Griffin-Novick	***Animal Structure and Function,*** *second edition*
Novikoff-Holtzman	***The Cell and Its Organelles***
Savage	***Evolution,*** *second edition*
Sistrom	***Microbial Life,*** *second edition*

Genetics

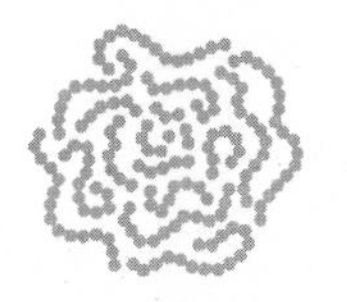

second edition

Robert Paul Levine

Harvard University

Holt, Rinehart and Winston, Inc.
New York Chicago San Francisco
Atlanta Dallas Montreal
Toronto London

Library of Congress Catalog Card Number: 68–20101

ISBN 0-03-068975-9

Printed in the United States of America

67890 006 98

Text and cover design by Margaret O. Tsao
Illustrations by George V. Kelvin • Science Graphics

Preface

The purpose of *Genetics*, second edition, is to convey to students who are beginning their study of biology a sense of the central role that both classical and molecular genetics play in all areas of biology. In order to understand this role a student must become familiar with certain fundamental facts and principles of genetics that apply to all forms of life from viruses to man. He must become acquainted not only with the classical patterns of inheritance that date from the discoveries of Mendel in 1866 but also with the present-day findings of molecular genetics that stem from the discovery in 1953 of the structure of DNA by Watson and Crick.

Three topics are emphasized in this book; namely, the nature of the genetic material, the transmission of the genetic material, and the function of the genetic material. Each of these topics contains information that is relevant to understanding many basic phenomena of biochemistry, physiology, and evolution, and a knowledge of each topic is a prerequisite for advanced studies of genetics.

Though *Genetics* was written with the beginning student in mind, it can serve as a brief review of the field for students who have some background in elementary genetics but who wish to bring their knowledge of the field up to date.

The preparation of the second edition of *Genetics* was aided greatly by the suggestions, criticisms and help of several people. I am indebted to the late Sir Ronald Fisher and to Dr. Frank Yates, F.R.S., Rothamsted, also to Messrs. Oliver & Boyd Ltd., Edinburgh, for permission to reprint Table III from their book, *Statistical Tables for Biological, Agricultural and Medical Research.* In particular, I wish to acknowledge the important contributions made by W. T. Ebersold, Ursula Johnson, N. W. Gillham, P. J. Hastings, Vicki Sato, and Elizabeth Levine.

R. P. L.
Cambridge, Massachusetts
March 1968

Contents

Genetics

Introduction

The dominant theme in the biology of all forms of life is a capacity for reproduction and, in viruses as well as men, this capacity leads to the transfer of hereditary characteristics from one generation to the next. The science of genetics is devoted to the study of hereditary characteristics and to three very general questions. *First,* what is inherited? What is the chemical and physical nature of the genetic material parents transmit to their offspring? *Second,* how is this genetic material transferred from parents to offspring? What are the mechanisms that bridge the gap between generations? *Third,* how does the genetic material act? Just as these three questions form the basis of the science of genetics, they also form the three main parts of this book; namely, the consideration of what is inherited, how it is transmitted, and finally, how it acts.

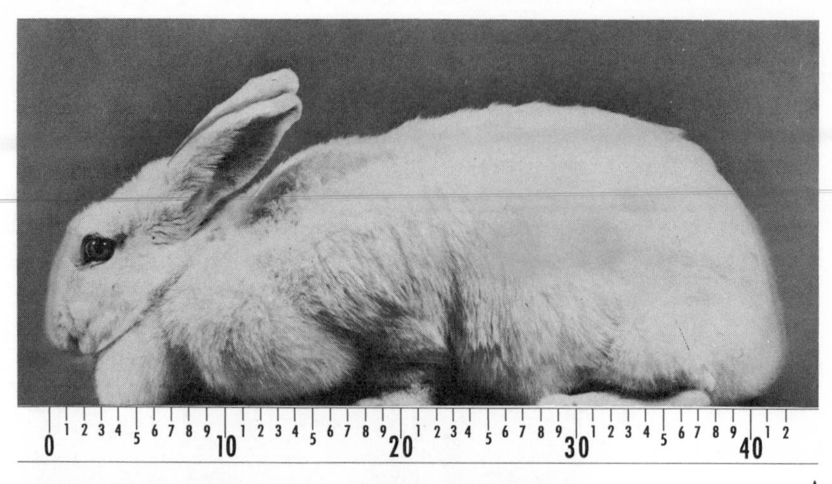

A

B

C

Phenotypic differences between three rabbits: (A) New Zealand white, (B) Polish white, and (C) Polish black. Their size is measured in centimeters. (Photographs by R. Stafford.)

THE PHENOTYPE An initial understanding of genetics comes from a recognition of what is meant by the inherent characters of an organism. The numerous features by which we recognize an organism constitute its *phenotype*. A rabbit, for example, possesses a combination of features by which we know it as a rabbit and thus distinguish it from other animals. Some of its *phenotypic characteristics* are the length of its ears, the short tail, and, in fact, the over-all conformation of its body shape. In addition, it is seen that there are differences among mature rabbits, as for instance, those shown on page 2. All of the animals in the photographs are rabbits, but there are certain *phenotypic differences* between them. Two of the rabbits are albinos, having no pigment in either their fur or eyes. In this respect they have the same phenotype. However, it is quite evident that they have different phenotypes for body size. The colored rabbit depicted in the same illustration is phenotypically distinct because its hair and eyes are pigmented; its phenotype for body size, however, is identical to that of the small albino rabbit. Thus, phenotypes are the means of recognition of similarities and differences between organisms.

A rabbit is phenotypically different from a field mouse in a number of respects. Even so, the two animals have many phenotypic similarities. They are both warm-blooded, vertebrate animals with central nervous systems and herbivorous feeding habits. The phenotype of the rabbit or mouse is to be contrasted with that of a garter snake, a vertebrate with a central nervous system, but cold-blooded and carnivorous. In these last two characteristics the garter snake is phenotypically similar to a garden spider, but the garden spider is an invertebrate and has a different kind of skeleton. Phenotypically distinct from these and many other multicellular animals are the numerous unicellular protozoans, and the vast array of lower and higher plants.

In spite of the many phenotypic differences among organisms, all are phenotypically similar in the myriad of biochemical processes necessary for life. This is a reflection of similarities at a molecular level, for the molecules of which they are made are essentially alike. In fact, molecular similarities extend to the viruses that infect animal, plant, and bacterial cells. The range of phenotypic characteristics, therefore, is a broad one and exists on many different levels.

The morphological or structural features of an organism constitute part of this range, and have their origin at the molecular level in the structure of macromolecules such as proteins and nucleic acids. At another level, morphological and structural phenotypes are seen under the electron microscope as similarities and differences in cell walls, in the fine-structural details within nuclei, and in other bodies within a cell. Yet another level is the microscopic anatomy of the cells as seen with the light microscope, where nuclei and chromosomes, for example, have characteristic phenotypes. Finally, there is the gross and microanatomy of tissues and organs that result in a phenotype comprising the entire façade of the organism.

In addition to the numerous morphological aspects of the phenotype, there is an array of biochemical and biophysical processes occurring at a molecular level—for example, the metabolism of glucose, the fixation of carbon dioxide during photosynthesis, and the transmission of electrical impulses along nerve fibers. These chemical and physical processes ultimately determine the pattern of reaction that an organism presents to its environment and, as such, are part of its phenotype.

At still another level are the physiological characteristics of the phenotype that result from the underlying biochemical and biophysical processes. These are exemplified by muscular contraction, circulation, development of secondary sex characteristics in animals, flowering time with respect to day length in plants, or the deciduous character of the leaves of some trees. There are, as well, phenotypic characteristics of behavior based on the physiology of the organism and its response to external stimuli.

Finally, one of the most important phenotypic characteristics of an organism is its capacity for reproduction. This begins with the duplication of macromolecules of nucleic acid, followed by the duplication of cell components such as chromosomes, chloroplasts, and mitochondria, and ultimately by the reproduction of the cells themselves. In turn, the means of cell reproduction—mitosis and binary fission in unicellular animals, mitotic reproduction of somatic cells in multicellular organisms, and meiosis in sexually reproducing unicellular and multicellular organisms—can be considered as part of the phenotype.

The phenotype is, therefore, many things. It may be an individual characteristic, such as coat color in a rabbit or the synthesis of a particular kind of molecule. On the other hand, it may be the combination of features that together contribute to the character of the organism as a whole.

THE GENOTYPE

Numerous phenotypic traits, such as coat color in rabbits, appear to be transmitted from one generation to the next. However, offspring do not inherit phenotypes from their parents; rather, they inherit the ability to produce these phenotypes. This ability resides in the *genotype*, and it is the material of the genotype that is transmitted from one generation to the next. The genotype is composed of numerous subunits called *genes*, which have specific chemical and physical properties that ultimately determine the nature of the phenotype. Each gene (or, more properly, the material of which it is composed) has the ability to reproduce itself exactly, and only rarely does this reproduction lead to a gene with properties different from that of the original. Thus there is continuity of the genotype from one generation to the next.

The expression of the phenotype is attributable not only to the genotype but to environmental conditions as well. These environmental conditions, however, do not affect the genotype. For example, a genotypic or genetic characteristic of most plants is their ability to synthesize chlorophyll in the light. If a plant such as corn is placed in the dark, chlorophyll synthesis will cease and the plant will become an albino. However, if it is put into the light it will once again synthesize chlorophyll. The effect of a dark environment upon this phenotype is thus temporary. Similarly, in man, the tanning of the skin occurs after exposure to ultraviolet irradiation, either naturally from the sun, or artificially with the aid of a sun lamp. The effect, however, is transient, for once the opportunity for exposure to ultraviolet light is removed, the skin becomes lighter again. Environmental alterations of the phenotype do not reflect alterations in the genotype, but rather the response of the organism to its environment. Indeed, the ability of an organism to respond to its environment is determined to a large extent by its genotype. The tanning of the skin in man is an example, for there are genetically different individuals who lack the ability to undergo the chemical reaction that leads to a tanned skin.

The environment, therefore, provides the arena in which the genotype acts; and accordingly, the phenotype represents the ultimate expression of the interaction of the genotype and its environment.

In the chapters that follow, a great deal of emphasis is placed upon the physical and chemical nature of the genetic material that comprises the genotype. The mode of transmission of the genetic material, as well as its mode of action, are interpreted on the basis of a knowledge of its physical and chemical properties. The three questions raised at the outset of this chapter will be answered to a large extent, but the answers themselves will have raised many new and intriguing questions regarding the genetic material and what it does. Indeed, genetics remains a subject that constantly rewards the investigator with stimulating and occasionally perplexing problems, all of which present an exciting challenge for scientific research.

part I

THE NATURE OF THE GENETIC MATERIAL

chapter **1**

The Structure and Duplication of the Genetic Material

A chemical analysis of an organism will reveal a variety of facts about its composition. Of the elements present, thirty-five are relatively common, and among the most common of these are carbon, hydrogen, oxygen, nitrogen, phosphorus, and sulfur. Certain of these elements are found in different classes of organic molecules such as fats, carbohydrates, lipids, vitamins, amino acids, purines, and pyrimidines. Many of these molecules are, in turn, formed into larger units called *macromolecules*. Proteins, for example, are long chains of amino acids held together by peptide bonds. They serve a number of important biological functions. For instance, proteins form the structure of skin, hair, and cartilage. In addition, a large number of proteins function as enzymes in catalyzing a multitude of biochemical reactions in all living organisms. From time to time enzymes will be mentioned in this book; the details of certain of their properties are covered in *Cell Structure and Function* by Ariel G. Loewy and Philip Siekevitz.

In addition to protein macromolecules, the class of macromolecules called *nucleic acids* is of special interest to geneticists and to biologists in general. Macromolecules of nucleic acid are composed of a five-carbon sugar (either deoxyribose or ribose), phosphorus, and several different nitrogen-containing organic molecules known as purines and pyrimidines. There are two kinds of nucleic acids, which differ in their chemical composition and structure. One of them is called *deoxyribonucleic acid*, or DNA, and the other *ribonucleic acid*, or RNA.

THE NUCLEIC ACIDS AS THE GENETIC MATERIAL

The nucleic acids were discovered in 1897 by F. Meischer, but their biological and genetic significance did not become fully appreciated until some 50 years later. In the intervening years F. Griffith, working with the bacterium which causes pneumonia, *Diplococcus pneumoniae*, performed experiments which were to lead to the identification of the nucleic acids as the genetic material.

Diplococcus pneumoniae, or pneumococcus for short, can exist in two different phenotypes known as smooth (*S*) and rough (*R*). *S* cells are *virulent* and are covered with a capsule of carbohydrate called a polysaccharide. *R* cells are *nonvirulent* and have no capsule. The various types of *S* bacteria (such as Type II*S* or III*S*) can be distinguished by differences in the chemical composition of the polysaccharide capsule. Each of these types is inherited, since the parent bacteria will reproduce their specific capsule type through countless cell generations. The ability to produce a specific type of polysaccharide capsule, then, is part of the organism's genotype. A given *S* type of pneumococcus can undergo an event known as *mutation* to yield a rare genetic variant or *mutant*, as it is called. About one *S* cell in ten million (or 10^7) undergoes a mutation to give rise to a colony composed of *R* cells. The *R* type is also inherited, for it is reproduced in subsequent generations. Moreover, a culture of these *R* cells can, in turn, give rise to an occasional *S* cell by a second mutation. When this occurs the *S* type is found to be identical to that of the smooth colony from which the original mutant *R* cell had been obtained.

Griffith's experiment utilized both *S* and *R* bacteria. Mice were injected with a *small number of living, nonvirulent R* pneumococci derived originally from a culture of Type II*S* bacteria. At the same time these mice also received an injection of a *large number of heat-killed* (and therefore no longer virulent) Type III*S* bacteria. Surprisingly, many of the mice died of pneumonia. Blood samples from the diseased mice showed the presence not only of *R* bacteria but also of large numbers of *virulent, encapsulated S bacteria*. These bacteria were found to be Type III*S* (Fig. 1-1). They could not have arisen by mutation, since the *R* bacteria were derived from Type II*S* and a mutation of these *R* bacteria gives rise only to Type II*S* bacteria. Thus, of the living bacteria injected into the mice—nonvirulent *R* bacteria originally derived from Type II*S*—some appeared to have been *transformed* into virulent Type III*S* bacteria. Furthermore, these newly formed Type III*S* bacteria could be shown to reproduce their type through many generations.

Later, several investigators confirmed Griffith's results, and among the most interesting experiments were those in which the transformation of *R* cells into *S* cells took place *in vitro*. For example, *R* cells were grown in a test tube in the presence of heat-killed *S* cells of a given type. Live *S* cells were found later among the *R* cells in the test-tube culture, and they were the same type as the heat-killed *S* cells (Fig. 1-2). In

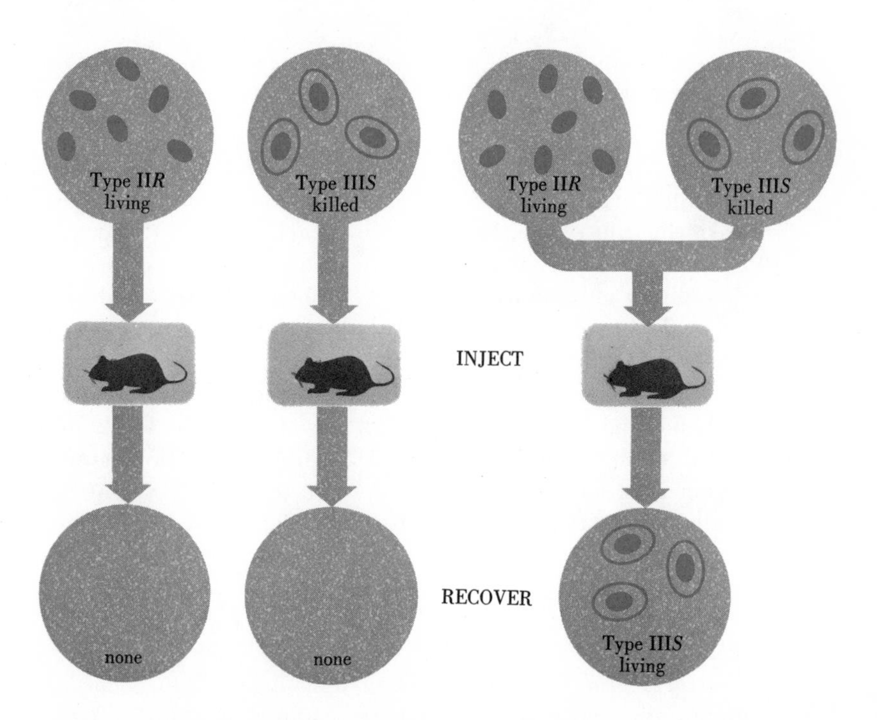

Fig. 1-1 *Diagrammatic representation of* in vivo *pneumococcus transformation experiment with mice. (Redrawn from Sutton,* Genes, Enzymes and Inherited Diseases. *New York: Holt, Rinehart and Winston, 1961.)*

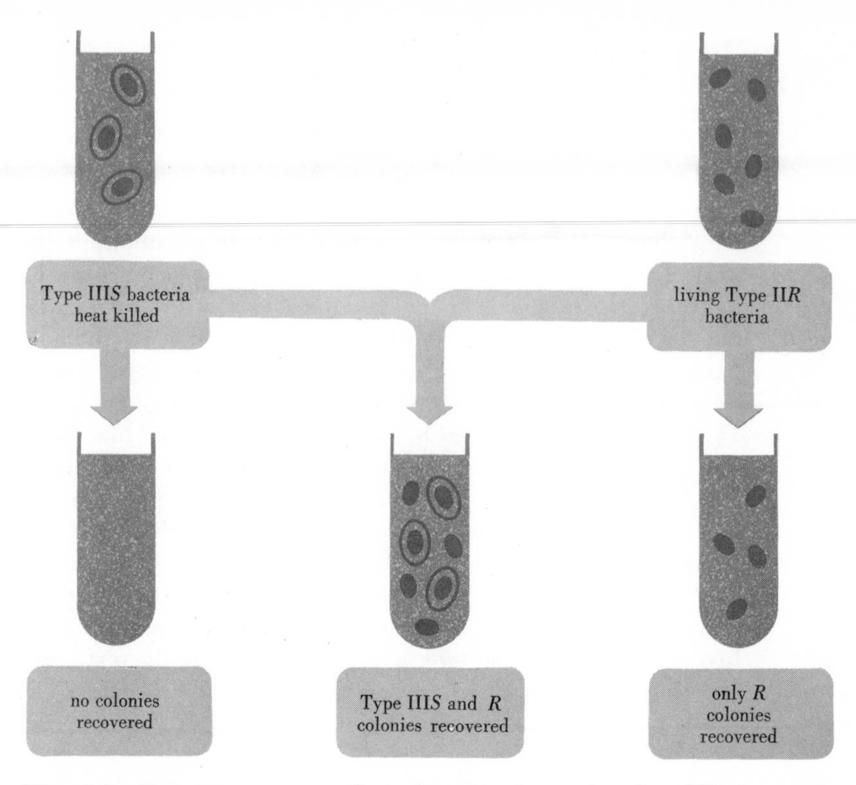

Fig. 1-2 *Pneumococcus transformation,* in vitro *using heat-killed smooth cells (diagrammatic).*

another *in vitro* experiment it was found that the presence of intact, heat-killed *S* cells was not necessary, and that some *R* cells could be transformed in the presence of *an extract* of *S* cells. There appeared, therefore, to be a substance in the *S* cells capable of bringing about a hereditary alteration of *R* cells. The next task was to determine the chemical nature of this substance—or *transforming principle,* as it came to be called.

The transforming principle, identified by O. T. Avery, C. M. MacLeod, and M. McCarty in 1944, was shown to be DNA. In the presence of highly purified extracts of DNA from Type III*S* bacteria, a small number of *R* bacteria were transformed to Type III*S*. However, if the *R* cells were exposed to DNA that had been treated with deoxyribonuclease, an enzyme that causes the destruction of DNA, no transformation took place (Fig. 1-3).

Experiments similar to those done by Avery, MacLeod, and McCarty have been repeated many times, and the phenomenon of *bacterial transformation* has been demonstrated to occur in several different species of bacteria. In each instance the transformation occurs when bacteria of one genotype (for example, *R* pneumococcus) are the recipients of DNA extracted from bacteria of a different genotype (such

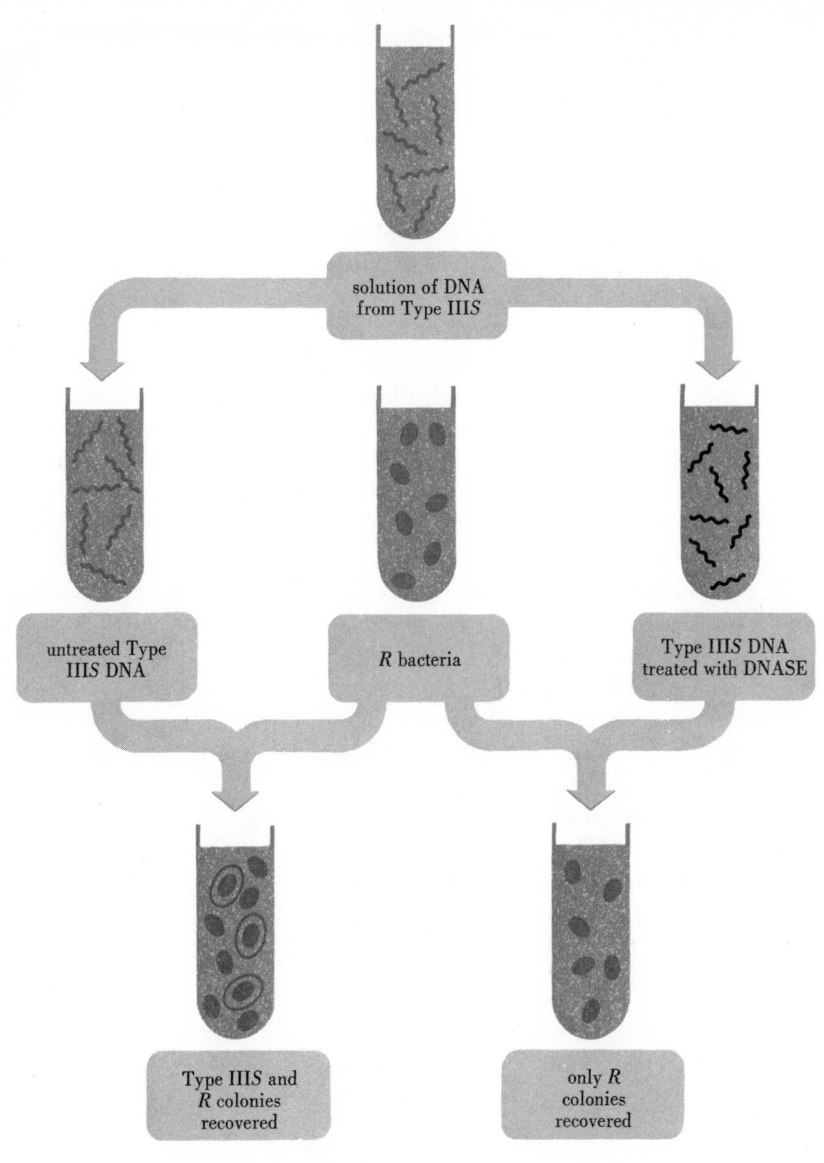

Fig. 1-3 *Transformation of pneumococcus using DNA extracted from heat-killed smooth cells.*

as *S* pneumococcus). Also, as in Griffith's original experiments, the cells that have been transformed have undergone a stable alteration in their genotype, and the new phenotype is expressed through many generations.

Transformation will be discussed again in Chapter 8, where it will be considered along with other aspects of the genetics of bacteria. At this point, it is important to understand that specific hereditary changes, or alterations in the genotype, can be brought about in pneumococcus and certain other bacteria when macromolecules of DNA from cells of one genotype are incorporated by cells of another genotype. Other macromolecules, such as proteins extracted from the bacteria, have no such transforming activity. It appears, therefore, that DNA has some sort of specific genetic activity and that it is the genetic material.

Additional support for this idea has been obtained from experiments with bacterial viruses or *bacteriophages.* There are several different types of bacteriophages (or phages for short) that infect the colon bacterium, *Escherichia coli.* Electron microscope studies show that the phages of *E. coli* have two major parts which have been designated as the *head* and *tail* (Fig. 1-4). The head and tail structures are protein and form the coat of the virus. DNA is found only within the head.

The initial step in phage infection is the attachment or adsorption of the tail of the phage to the bacterial cell surface. Then, in a matter of minutes, phages are reproduced within the bacteria; the infected bacteria break open, or *lyse*; and hundreds of the new phage progeny are released. The protein and the DNA of these new phages are formed from constituents of the bacterial protoplasm, ultimately killing the host bacteria. As discussed later (Chapter 8), phages can differ in a number of genetic characteristics, and when a bacterium is infected with a single phage of a given genotype, the progeny that emerge upon lysis of the bacterium are genetically identical to the original infecting phage.

Of interest at this point is whether the phage DNA or the phage protein, or perhaps both, are necessary for the process of phage reproduction. In 1952, A. Hershey and M. Chase performed an ingenious experiment to answer this question (Fig. 1-5).

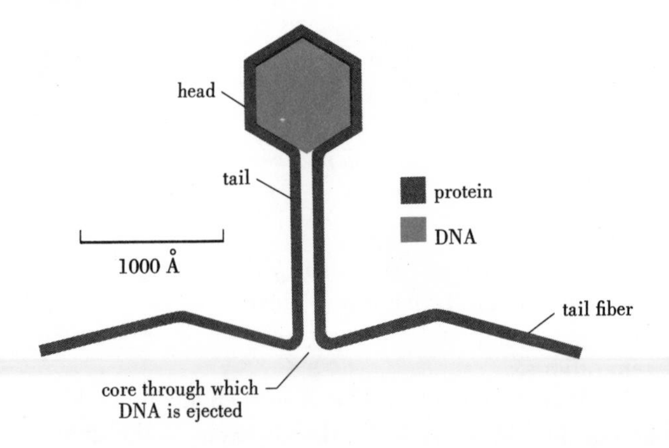

Fig. 1-4 Diagram of bacteriophage. (Redrawn from Sutton, Genes, Enzymes and Inherited Diseases. *New York: Holt, Rinehart and Winston, 1961.)*

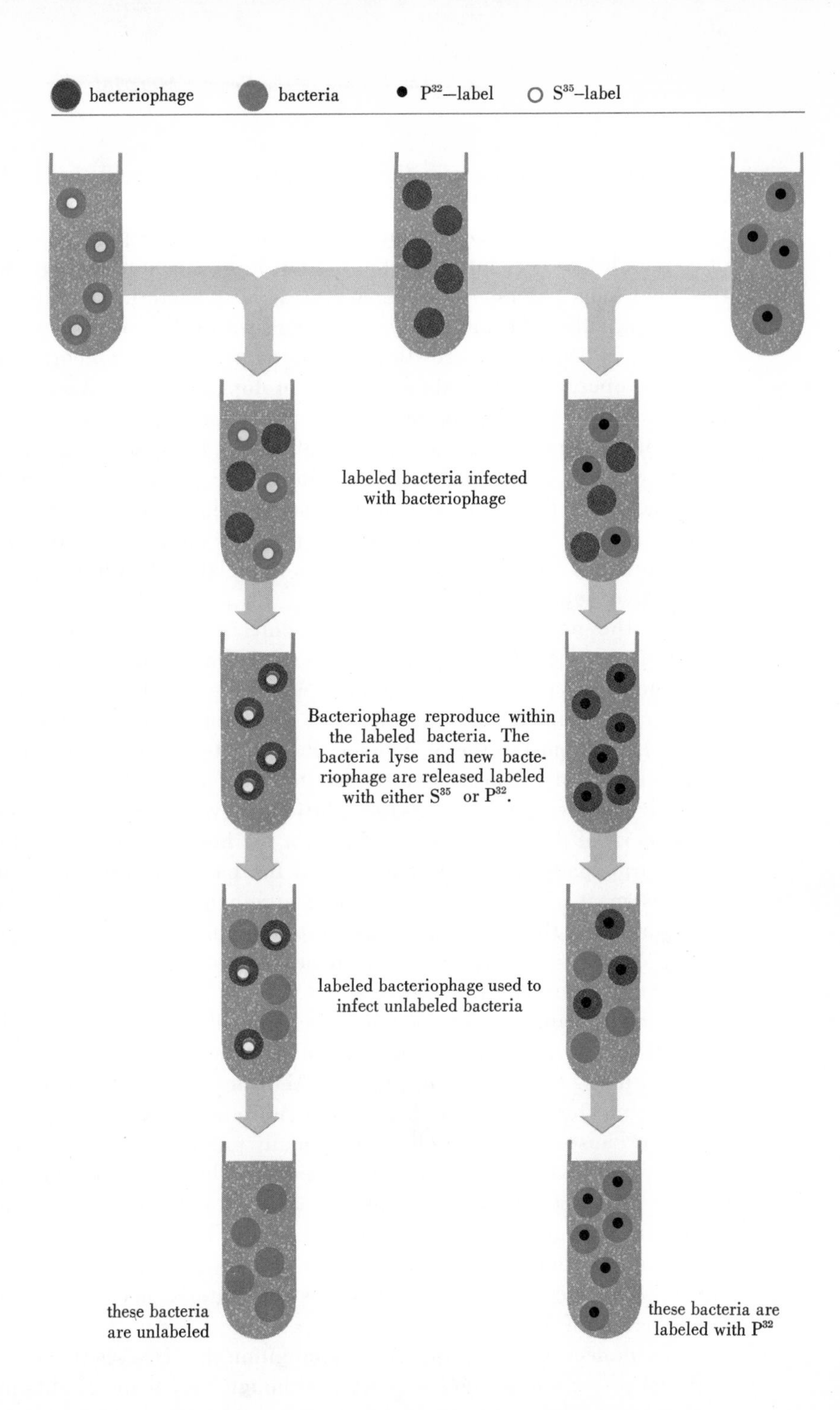

Fig. 1-5 *Diagrammatic representation of the Hershey and Chase experiment with isotopically labeled bacteriophages (see text for details).*

Analysis of the composition of the phages had revealed that they were composed primarily of protein and DNA. Phosphorus is one of the major constituents of DNA, but sulfur is absent. The protein, on the other hand, has very little phosphorus but it does contain sulfur. With this knowledge, Hershey and Chase grew host bacteria in a medium containing either the radioactive isotope of sulfur (S^{35}) or that of phosphorus (P^{32}). During the growth process, the host bacteria incorporated these isotopes, thereby "labeling" their protoplasmic constituents with either radioactive sulfur or radioactive phosphorus. (The two isotopes were not combined in the same bacterial culture because it was difficult to differentiate between them when both were present.) The labeled bacteria were then infected with phages. The phages reproduced within the bacteria and after lysis the progeny phages were released. These progeny were collected and found to be labeled with the specific radioactive isotope of the host bacteria.

In the next step of the experiment, unlabeled bacteria were infected with the labeled phage, and the distribution of the label in the host bacterium was then determined. When the infection was brought about by S^{35}-labeled phages, little label was found within the host bacteria. Instead, most of the S^{35}-labeled phage protein was found attached to the outside of the host bacteria, in the coats of the infecting viruses. When P^{32}-labeled phages were used, little label was found in the protein coats; most of the label was found *within* the host bacteria. Therefore, it was principally the DNA that entered the bacteria during infection, whereas most of the protein remained attached to the outside of the bacteria. Thus, the material injected into the bacterium by the virus is DNA, and it is this DNA that is necessary for the reproduction of genetically identical virus particles.

Similar evidence shows that a second nucleic acid, *ribonucleic acid* (RNA), is the genetic material of certain plant viruses. One of these viruses, *tobacco mosaic virus* (TMV), infects the tobacco plant by entering the cells of the leaves and reproducing within them. The infection causes yellow patches on normally green leaves. Like phage, TMV is composed of protein and nucleic acid, but in this instance the nucleic acid is RNA. The protein and RNA of TMV have been separated, purified, and tested for their infectivity. It has been found that the RNA is capable of infecting the host plant, but the protein is not. Again, a nucleic acid, this time RNA, is the material necessary for the reproduction of new virus particles.

Evidence exists to support the contention that DNA is the genetic material of higher organisms as well, although the means of obtaining such evidence has not been as direct as in the bacteria and the viruses. One feature of the genotype is its transmission, essentially unchanged, from one generation to the next. The implication is that the genetic material is transmitted unchanged from generation to generation. Therefore, one of the characteristics of the genetic material should be its constancy from cell to cell and from generation to generation. In other words, the genetic material should not be metabolized. Experiments with different organisms have demonstrated that although proteins and carbohydrates, for example, undergo continual breakdown and resynthesis in cellular metabolism, DNA does not. Measurements of the amount of DNA per cell of a given organism show that it remains essentially constant, whereas the amounts of other cell constituents vary from one cell type to another.

The DNA of many bacteria and of all higher organisms occupies specific sites within the cell. In higher organisms the principal site is the cell nucleus. Relatively small amounts of DNA have also been found outside the nucleus in mitochondria and chloroplasts, but special techniques are required for DNA detection in these structures (see Chapter 7). Nuclear DNA, however, can be detected easily by the application of certain stains which produce specific color reactions. One of these stains, the *Feulgen* stain, reacts specifically with the deoxyribose sugar of DNA. Within the nucleus the staining reaction occurs in long, threadlike structures called *chromosomes.* The DNA, therefore, is localized in the chromosomes within the cell nucleus.

The amount of DNA in a nucleus can be determined in several ways. One method is to extract the DNA from the cell by chemical means and to purify it. You will recall that this was done for pneumococcus DNA. If the initial number of cells undergoing extraction, or their dry weight, or some other measure of their number is known, the average amount of DNA per cell can then be determined. In another method, the amount of DNA can be ascertained by measuring the extent of the Feulgen reaction within the nucleus by the use of sensitive photometric equipment that responds to the amount of light transmitted through the stained nucleus. The amount of DNA in unstained nuclei can also be determined photometrically. In this method, the amount of light of a specific wavelength absorbed by DNA is measured. Both chemical analysis and photometric determinations of stained and unstained nuclei show that the amount of DNA is the same in different kinds of cells from the same organism.

The amount of DNA in a cell changes, however, during chromosome duplication prior to cell division. When the chromosomes duplicate, the amount of DNA doubles. After the cell has divided, each of the two daughter cells once again contains the original amount of DNA. This amount remains constant until their chromosomes duplicate in anticipation of the next cell division. The conclusion reached from the measurement of the amount of DNA per cell is that it is constant except when chromosomes duplicate.

THE STRUCTURE AND COMPOSITION OF THE GENETIC MATERIAL

Earlier in this chapter it was mentioned that both DNA and RNA are composed of the nitrogen-containing bases, purines and pyrimidines, a five-carbon sugar, and phosphate. There are, however, two major differences between DNA and RNA. In DNA the five-carbon sugar is deoxyribose whereas in RNA it is ribose (Fig. 1-6). In addition, the pyrimidines of DNA and RNA differ (Fig. 1-7). In DNA the purines are *adenine* and *guanine* and the pyrimidines are *thymine* and *cytosine*. Adenine, guanine, and cytosine are also present in RNA, but the pyrimidine *uracil* is found instead of thymine. In the DNA of some organisms cytosine can be replaced by closely related molecules (such as 5-hydroxymethyl cytosine). A small number of other bases can also be found in RNA, but the units that make up DNA or RNA are similar in all organisms.

Both DNA and RNA are comprised of units called *nucleotides*. A nucleotide is a compound consisting of a purine or a pyrimidine and a phosphate group bound to a sugar molecule (Fig. 1-8). In both DNA and RNA the nucleotide units are linked to form a *polynucleotide chain* (Fig. 1-9). In this chain the sugar of one nucleotide is bound to the phosphate group of another nucleotide, and so on.

The three-dimensional structure of DNA was established by M. H. F. Wilkins, R. Franklin, and their co-workers in studies of the x-ray diffraction patterns produced by isolated DNA fibers. The data they obtained from such patterns is too complex to be presented in this discussion, but their interpretation of the data may be summarized. First, they found that DNA from several different species had identical spatial orientations; thus they could assume the existence of a constant DNA conformation in the cell. Second, they found that the DNA was *helical* in structure and that there were at least two helices. Since DNA is composed of polynucleotide chains, they proposed that two or more of these chains were helically coiled about each other to form the DNA macromolecule.

Chemical analysis revealed another important fact about DNA. E. Chargaff found a specific, quantitative relationship between the nucleotides obtained from the chemical breakdown of calf thymus

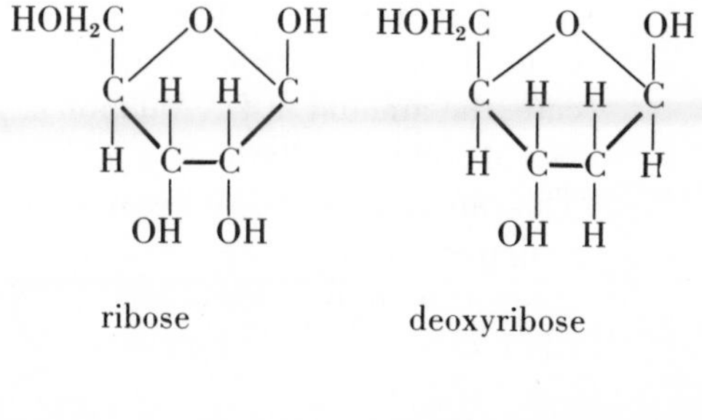

ribose

deoxyribose

Fig. 1-6 *The structures of ribose and of deoxyribose, the five carbon sugars (pentoses) of RNA and DNA respectively.*

Fig. 1-7 Purine and pyrimidine bases. Adenine, guanine, cytosine, and thymine occur in DNA. In RNA thymine is replaced by uracil.

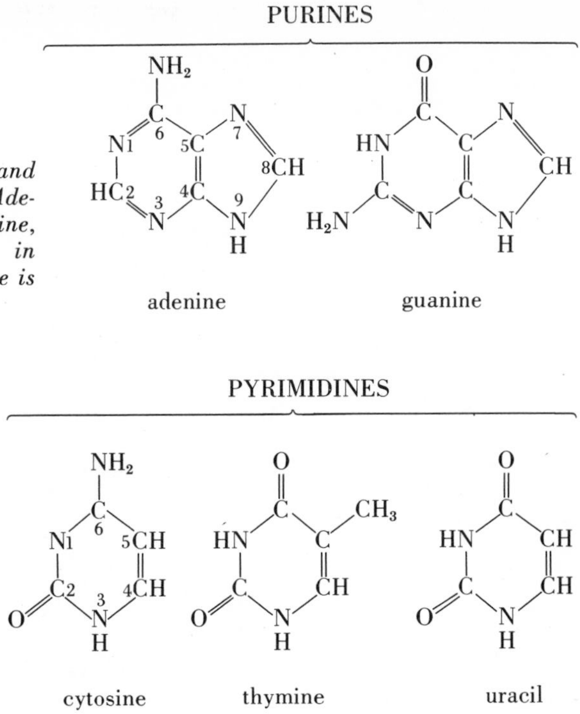

PURINES

adenine guanine

PYRIMIDINES

cytosine thymine uracil

DNA. For each nucleotide of adenine there is one of thymine (A = T), and for each nucleotide of guanine there is one of cytosine (G = C).

The data from x-ray diffraction studies, as well as from the chemical analyses of DNA, provided the information used by J. D. Watson and F. H. C. Crick in proposing a model for DNA structure. They made two assumptions in designing their model. First, they assumed that

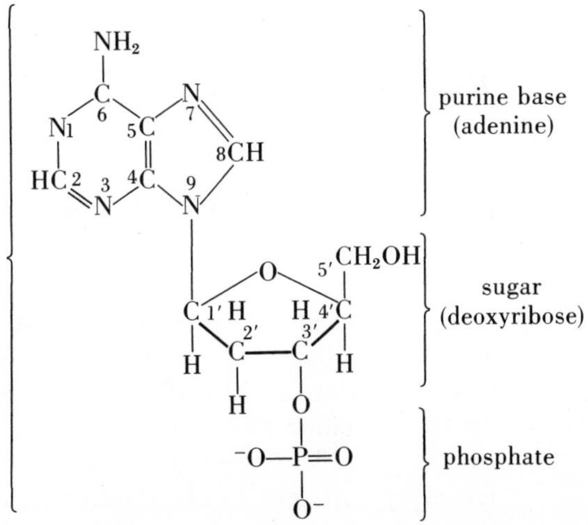

nucleotide (deoxyadenylic acid)

purine base (adenine)

sugar (deoxyribose)

phosphate

Fig. 1-8 A nucleotide.

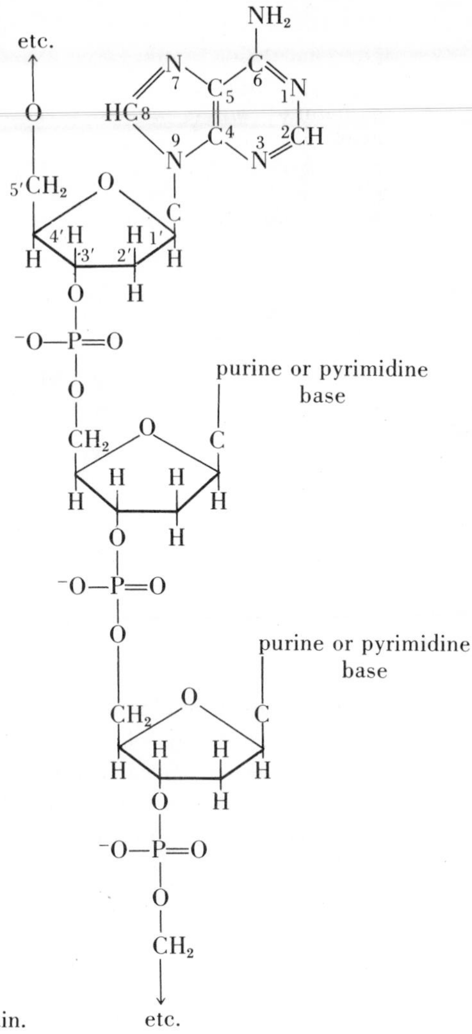

Fig. 1-9 A polynucleotide chain.

there were two helically coiled polynucleotide chains. The diffraction patterns, which revealed the three-dimensional structure of DNA, had suggested that there were at least two. Second, they assumed that the two polynucleotide chains were bound together in a specific fashion. This assumption was based on two pieces of experimental evidence. One of these was Chargaff's observation of the reciprocal relationship between adenine and thymine (A = T) and between guanine and cytosine (G = C). The other evidence was that a pair of nucleotides can be joined by *hydrogen bonds* (H bonds), and that the H bonding between the purines and the pyrimidines of these nucleotides is highly specific. Adenine, for example, forms two H bonds with thymine; guanine forms three H bonds with cytosine (Fig. 1-10). The formation of H bonds between adenine and cytosine, or adenine and guanine is physically very difficult, and requires a marked distortion of the helical structure of DNA. Similarly, guanine cannot easily pair with adenine or thymine. Such pairing of nucleotides would be highly unlikely.

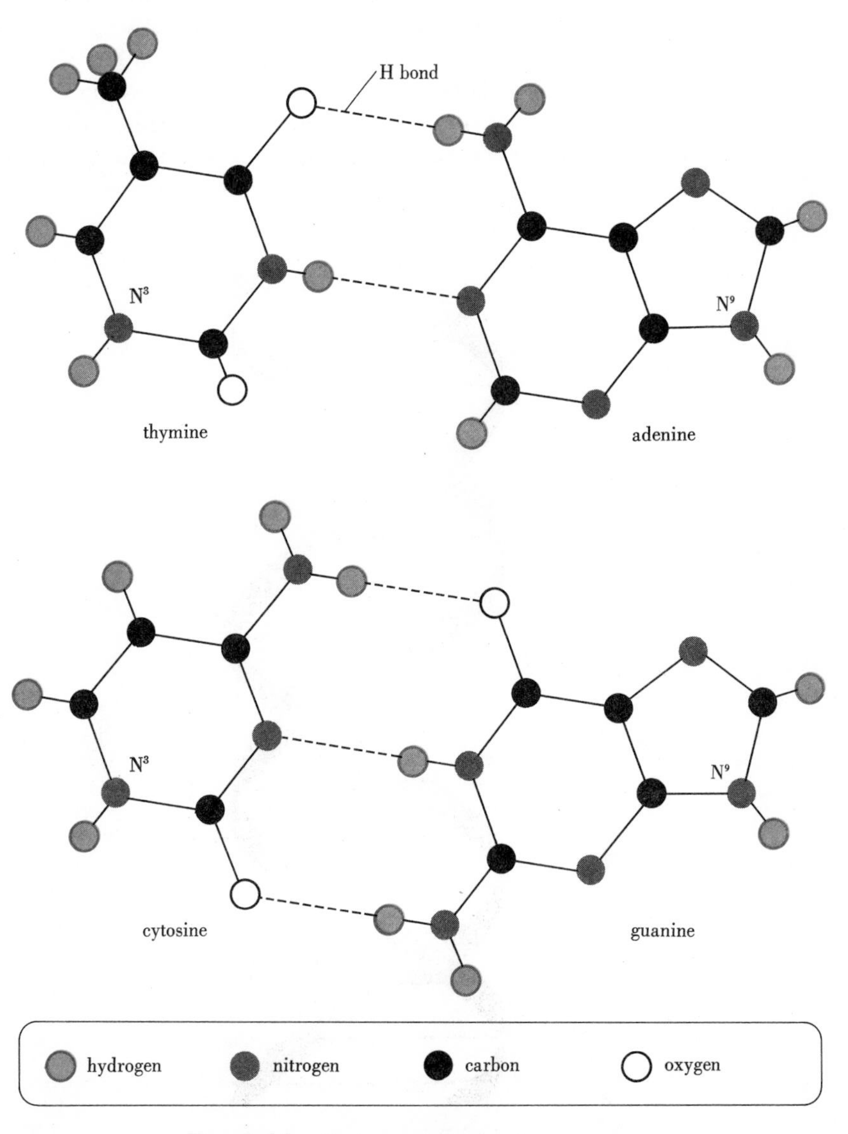

Fig. 1-10 *Hydrogen bonding between bases.*

Thus, Watson and Crick proposed that an individual polynucleotide chain consists of nucleotides linked lengthwise by way of sugar and phosphate groups. Two helically coiled polynucleotide chains are bound together crosswise by hydrogen bonding between a purine and a pyrimidine. The restriction on the pairing of nucleotides is imposed by the limitation of the H bonding of adenine to thymine and of guanine to cytosine.

The structure of DNA as proposed by Watson and Crick is shown in Figs. 1-11 and 1-12. In Fig. 1-11 the helical structure is shown in a diagrammatic fashion. Each helical strand represents a single polynucleotide chain. The H bonding that connects the two polynucleotide chains is indicated by the horizontal bars. The dimensions suggested from the x-ray diffraction patterns are given in angstroms (an angstrom, Å, is equal to 0.1 mμ or one ten-thousandth of a micron). Figure 1-12

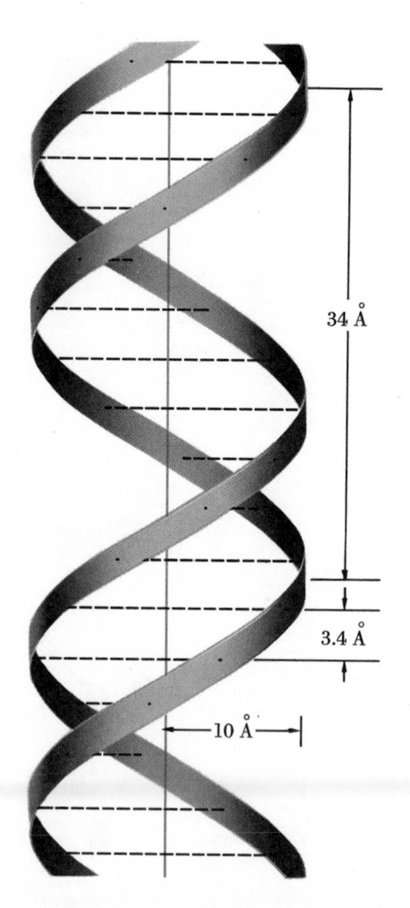

Fig. 1-11 The Watson-Crick DNA double helix. (After Watson and Crick, Cold Spring Harbor Symposia on Quantitative Biology, *Vol. 18, 1953.)*

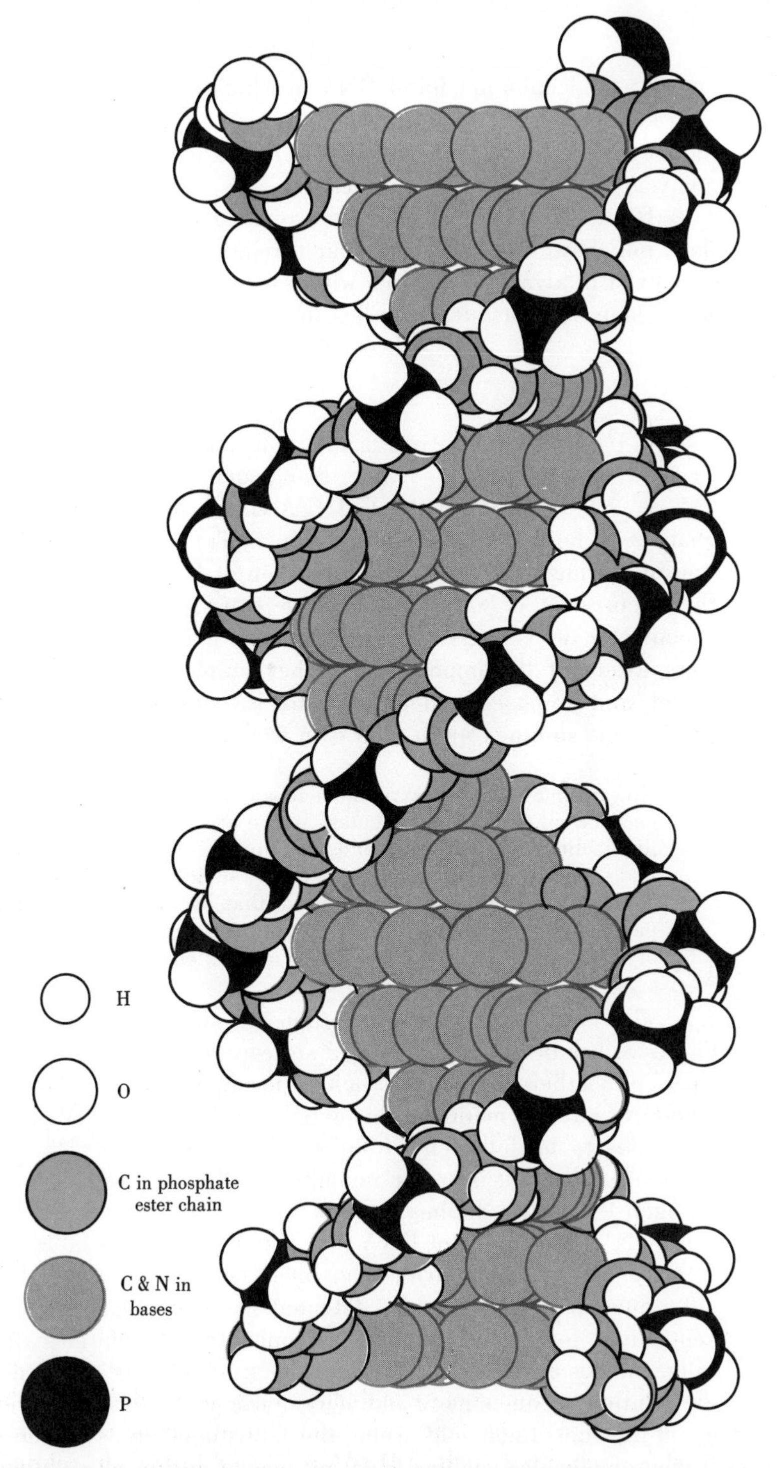

Fig. 1-12 *Molecular model of DNA. (After Feughelman,* et al., Nature, *Vol. 175, 1955. In Sager and Ryan,* Cell Heredity. *New York: Wiley, 1961.)*

shows a molecular model of DNA in which each atom is represented on a highly magnified but exact scale. It is important to realize that this model fits the three-dimensional structure originally proposed from the analysis of x-ray diffraction patterns.

RNA, like DNA, is composed of long, polynucleotide chains. It does not, however, have a regular three-dimensional structure and it may exist in a variety of forms within the cell (see Chapter 10). A complete analysis of its structure has thus not been accomplished to date.

THE DUPLICATION OF DNA

It is apparent that if the sequence of nucleotides along one polynucleotide chain is A G T C A C, the sequence along the other chain must be the complementary sequence of T C A G T G. The restriction in pairing imposed by the limitation of H bonding is the basis for the occurrence of *specific nucleotide pairs,* and such pairing could account for the conservation of the genetic material. Watson and Crick were aware of the importance of the complementary nature of their model, and proposed a mechanism for the duplication of DNA, which can be best summed up in their own words:

> Now our model for deoxyribonucleic acid is, in effect, a *pair* of templates, each of which is complementary to the other. We imagine that prior to duplication the hydrogen bonds are broken, and the two chains unwind and separate. Each chain then acts as a template for the formation on to itself of a new companion chain, so that eventually we shall have *two* pairs of chains, where we only had one before. Moreover, the sequence of the pairs of bases will have been duplicated exactly.

The mechanism that they proposed, shown diagrammatically in Fig. 1-13, has been supported by a series of experiments with bacteria, algae, and other organisms. Such experiments involve the use of isotopes. As already mentioned, isotopes can be used as labels, allowing molecules to be followed within a cell. DNA can be labeled with a number of different isotopes, including that of nitrogen. The isotope of nitrogen known as N^{15} has a mass that is greater than ordinary nitrogen or N^{14}. Molecules of DNA that contain an appreciable amount of N^{15} are denser and heavier than those containing only N^{14}, and this difference in density can be detected by centrifugation. When high centrifugal forces are applied to concentrated solutions of certain salts, such as cesium chloride, a density gradient is established in which the solution becomes more and more dense as it approaches the periphery of the centrifugal field, where the centrifugal forces are the greatest. If macromolecules such as DNA are placed in the salt solution, during

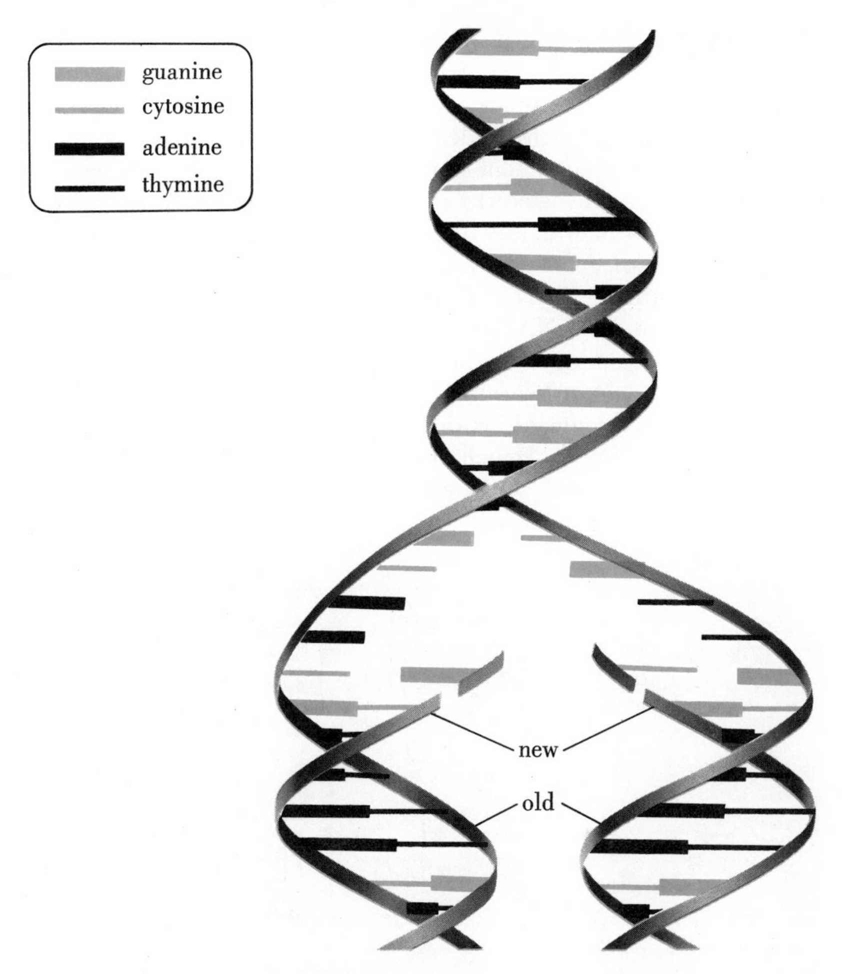

Fig. 1-13 The replication of DNA according to Watson and Crick. (Redrawn from Sutton, Genes, Enzymes and Inherited Diseases. *New York: Holt, Rinehart and Winston, 1961.)*

high-speed centrifugation they will reach a point in the gradient at which their density is equal to that of the salt solution. Since DNA containing N^{15} is denser than DNA with N^{14}, the two can be distinguished in the density gradient.

Experiments using N^{15}-labeled DNA were carried out in 1958 by M. Meselson and F. Stahl. *E. coli* were grown for several generations on a medium in which all of the nitrogen was N^{15}. The DNA extracted from these bacteria was fully labeled with the heavy isotope, and upon centrifugation it proved to be denser than normal DNA. Cells with fully labeled N^{15}-DNA were then transferred to a medium that contained only N^{14} as a nitrogen source. Next, the culture was sampled at various times to ascertain the density of the DNA. The sample times corresponded with the times of doubling of the cells within the culture. The first doubling corresponded to the first synthesis of DNA in the *absence* of N^{15}, the second doubling time to a second synthesis, and so on. The first sample revealed the presence of DNA with an intermediate density; that is, it was less dense than the original, fully labeled N^{15}-DNA present just before the bacteria reproduced in an N^{14} medium, but was denser than N^{14}-DNA. Thus, there was *hybrid*-DNA in contrast to N^{14}-DNA and N^{15}-DNA. After the second doubling of the DNA, and the second doubling of the cells, a sample revealed two kinds of DNA; an equal amount of hybrid DNA and N^{14}-DNA. A third sampling revealed the same amount of hybrid DNA as in the previous sample, but there was three times as much of the N^{14}-DNA present.

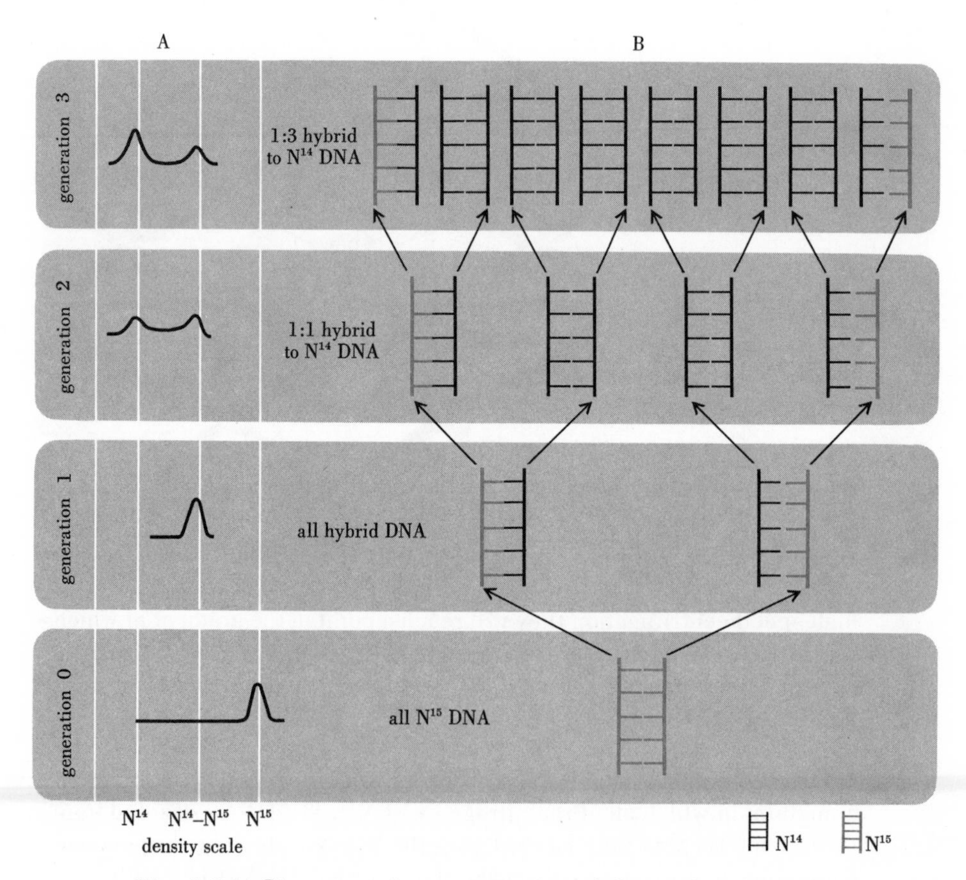

Fig. 1-14 Diagrammatic representation of the results of the Meselson and Stahl experiment on DNA replication (see text for details).

These results are shown diagrammatically in Fig. 1-14A, which indicates the distribution of the DNA densities due to the presence of N^{14} and N^{15}. If DNA were to duplicate by first undergoing breakdown into small fragments (that is, by some *dispersive* means), the first generation of DNA molecules would be comprised not of hybrid DNA but rather of an array of DNA's with densities ranging from N^{14} to N^{15}. If the *whole* DNA molecule were to act as a template for a new DNA molecule, the first generation would consist of the original N^{15}-DNA and new N^{14}-DNA. Again, hybrid DNA would be absent. Replication by this mechanism is called *conservative*, for the original DNA is conserved. The results obtained by Meselson and Stahl are not compatible with either of these mechanisms of duplication. Their results (Fig. 1-14A) are interpreted in the manner shown in Fig. 1-14B. It will be seen that there is a remarkable similarity between their interpretation and the scheme proposed by Watson and Crick (Fig. 1-13). The DNA replicates in a *semiconservative* manner; that is, the original strands are conserved but have separated and formed new strands upon themselves. Alternative models can be proposed from the Meselson and Stahl data, but the simplest interpretation of their results agrees with the Watson and Crick model for the duplication of DNA. Identical experimental results have been obtained in organisms other than *E. coli.* For example, the DNA of mitotically dividing cells of the green alga *Chlamydomonas reinhardi*, as well as other nucleate organisms, has been shown to duplicate semiconservatively. Recently, it has been shown for *Chlamydomonas* that the duplication of DNA during meiosis, and the duplication of chloroplast DNA, is semiconservative. Wherever tested, the Watson and Crick hypothesis for the mode of DNA duplication has been proved accurate.

Although the duplication of DNA has been shown to be semiconservative, the mechanism for the synthesis of the new strands is not yet fully understood. However, it is known that the synthesis of new DNA requires the presence of "old" DNA as a primer, an enzyme called DNA polymerase, and each of the four bases.

THE SIGNIFICANCE OF SPECIFIC PAIRING BETWEEN NUCLEOTIDES

There is, at present, little doubt that the genetic material is DNA (or RNA in certain viruses). The weight of evidence favors the structure for DNA proposed by Watson and Crick, and the implications of this structure for genetics are profound.

One of the most important properties of this structure of DNA lies in the restriction of pairing between nucleotides. This restriction ensures that one DNA molecule will give rise to two identical daughter molecules. In genetic terms this means that one cell will give rise to two daughter cells possessing identical genetic material, or, in other words, to two cells with identical genotypes.

DNA is the source of *genetic information* necessary for the development of phenotypes. Every organism possesses a vast array of hereditary characteristics, and these in turn reflect many different kinds of genetic information. Diversity of genetic information is found in DNA if DNA is assumed to be a *coded* molecule. Consider the four nucleotides as the only letters in an alphabet for a code or language that depends upon the sequence and perhaps the ratio of pairs of A — T and G — C nucleotides. A DNA molecule may contain 10^7 pairs of nucleotides. If we assume that half of the pairs are A — T and half are G — C, the number of different sequences of pairs possible is immense.

A discussion of the function of the genetic material and the genetic code must be delayed until later chapters. However, it will be seen at this point that the structure of DNA satisfies important requirements that must be met by the genetic material. In particular, the pairing restriction preserves identical arrays of nucleotide pairs, which are the basis for a genetic code or language.

FURTHER READING

Anfinson, C. B., 1959. *The molecular basis of evolution.* New York: John Wiley & Sons, Inc.

Ingram, V. W., 1965. *The biosynthesis of macromolecules.* New York: W. A. Benjamin, Inc.

Watson, J. D., 1965. *The molecular biology of the gene.* New York: W. A. Benjamin, Inc.

chapter 2

Chromosome Duplication and Division

The DNA contained within the cell nucleus is concentrated into highly organized structures known as *chromosomes*. During nuclear divisions the chromosomes go through a series of characteristic developmental stages known as *mitosis* and *meiosis*. Although these stages have not been demonstrated among nonnucleate organisms such as bacteria, it will be shown in Chapter 8 that their DNA is organized into chromosome or chromosome-like structures.

MITOSIS Unicellular organisms, such as protozoa and many algae, increase in number through simple fission. This asexual process of reproduction, whereby one cell gives rise to two identical daughter cells, follows the division of the nucleus known as *mitosis.* A multicellular organism develops from a fertilized egg or a *zygote* by mitosis. Furthermore, the constant replacement of cells, such as the epithelial cells of the skin or the erythrocytes of the blood, is by the same process.

The DNA of a cell is duplicated before the onset of mitosis so that each daughter cell will contain the same genetic information as the mother cell. Mitosis can be viewed as the mechanism whereby this duplicated set of identical genetic information, contained in chromosomes, is distributed equally to the nuclei of the two daughter cells.

The mitotic process can be divided into four stages, each characterized by certain features of nuclear or chromosome morphology and chromosome movement. The stages are arbitrary in that mitosis is a continuous process from its inception at *prophase*, through the stages of *metaphase* and *anaphase*, to the final stage of *telophase*. Figure 2-1 is a diagrammatic representation of mitosis and Fig. 2-2 is a series of photographs of mitosis in the peony.

The nucleus of a cell that is not dividing is in the *interphase* stage (Fig. 2-1A). The DNA within the interphase nucleus appears diffuse and evenly distributed; chromosomes are not visible. As mitosis begins, the nucleus undergoes marked alterations. A coiling process begins, resulting in the appearance of threadlike chromosomes. This initial stage of mitosis (Fig. 2-1B) is referred to as *prophase*. The condensation of the nuclear material continues throughout this stage until chromosomes are clearly visible. A microscopic examination of the prophase chromosomes reveals that from the inception of prophase each chromosome is comprised of two strands. Each strand is called a *chromatid*, and the two chromatids of the chromosome are known as *sister chromatids*. Sister chromatids are formed as a result of the DNA duplication which has occured during the previous interphase. Thus, two sister chromatids of a given chromosome contain identical genetic information.

Prophase of mitosis ends, and the next stage, *metaphase*, begins when the double membrane surrounding the nucleus disperses, the chromosomes become aligned along the equator of the cell, and the spindle begins to form (Fig. 2-1C). The spindle is composed in part of protein fibers called microtubules. These microtubules are believed to play a role in the movement of chromosomes during nuclear divisions. The sister chromatids of each chromosome are held together at a specialized region known as the *centromere*, which becomes associated with spindle fibers.

The next stage of mitosis, *anaphase* (Fig. 2-1D), is initiated when the centromere divides, and the sister chromatids separate. Each chromatid now moves, centromere first, to opposite poles of the spindle. Once the chromatids have moved to their respective poles they are considered as *daughter chromosomes*.

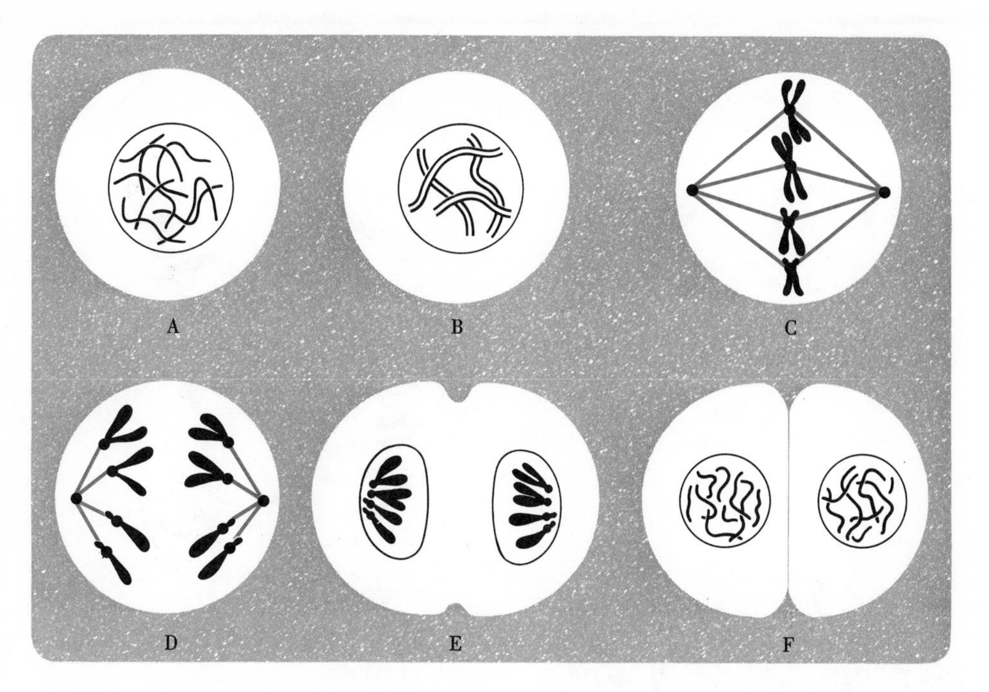

Fig. 2-1 Diagrammatic representation of the stages of mitosis (see text for details).

After the movement to the poles is complete and a double nuclear membrane begins to form around each set of daughter chromosomes, the stage of *telophase* has been reached (Fig. 2-1E). The chromosomes become less distinct again, and the nucleus returns to interphase (Fig. 2-1F). The division of the cell, as distinct from the nuclear division just described, proceeds after mitosis has taken place.

In summary, at the conclusion of mitosis, both daughter nuclei have an identical complement of chromosomes and, thus, of DNA. This results from the separation of the identical sister chromatids at anaphase, and the orderly distribution of the genetic material is, in turn, the consequence of the alignment of the identical sister chromatids along the equator of the cell at metaphase.

MEIOSIS

The growth of multicellular organisms, as well as asexual reproduction in unicellular ones, is a consequence of mitotic divisions. In sexual reproduction, however, *gametes* or *germ cells* arise by *meiosis* which consists of two successive nuclear divisions similar to mitosis. In contrast to mitosis, however, they are accompanied by *only one chromosomal division*. The final result, of course, is that the chromosome number is reduced by one half. When gametes fuse, the original chromosome number will be

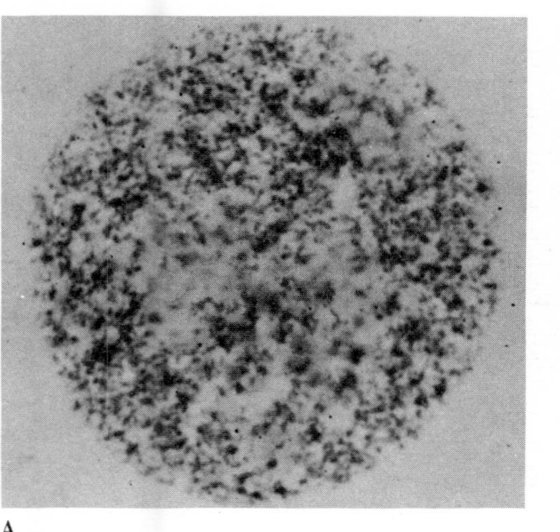
A

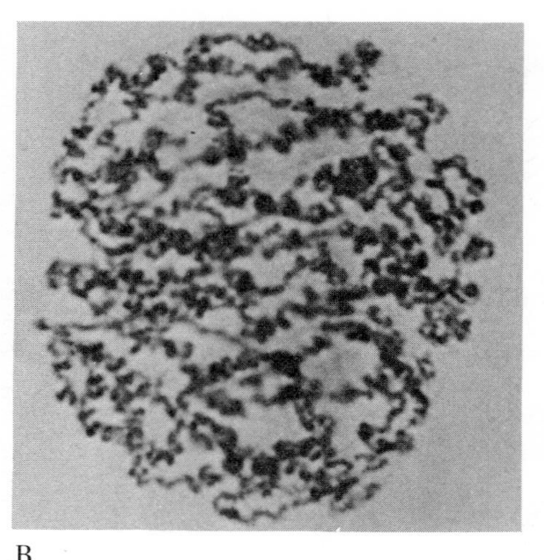
B

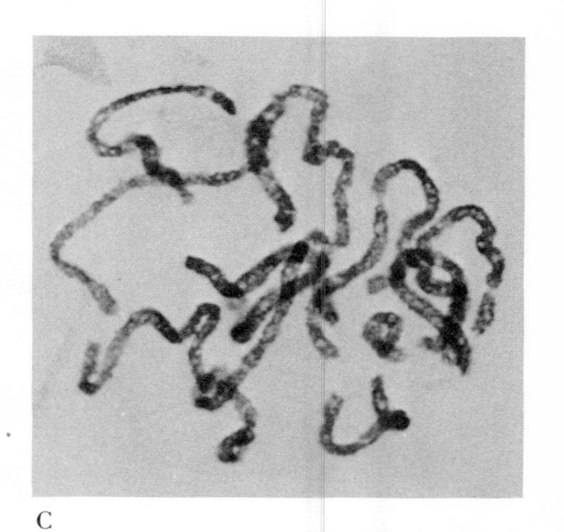
C

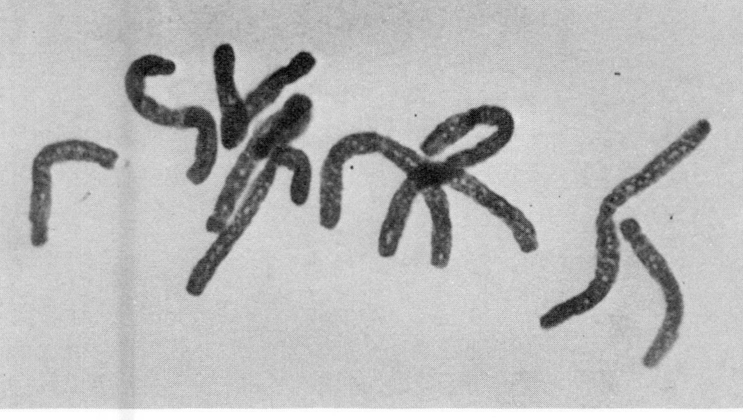
D

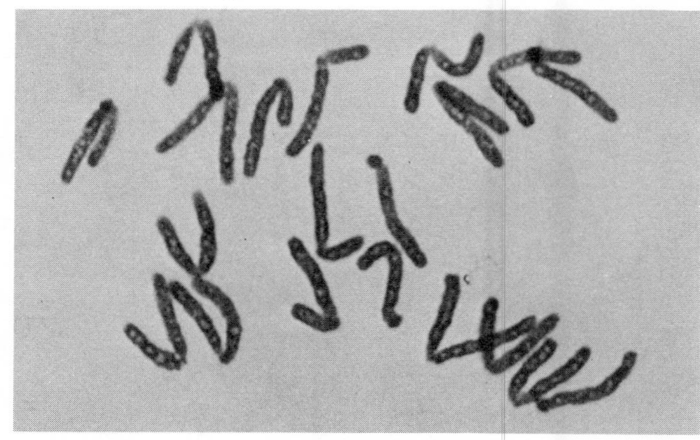
E

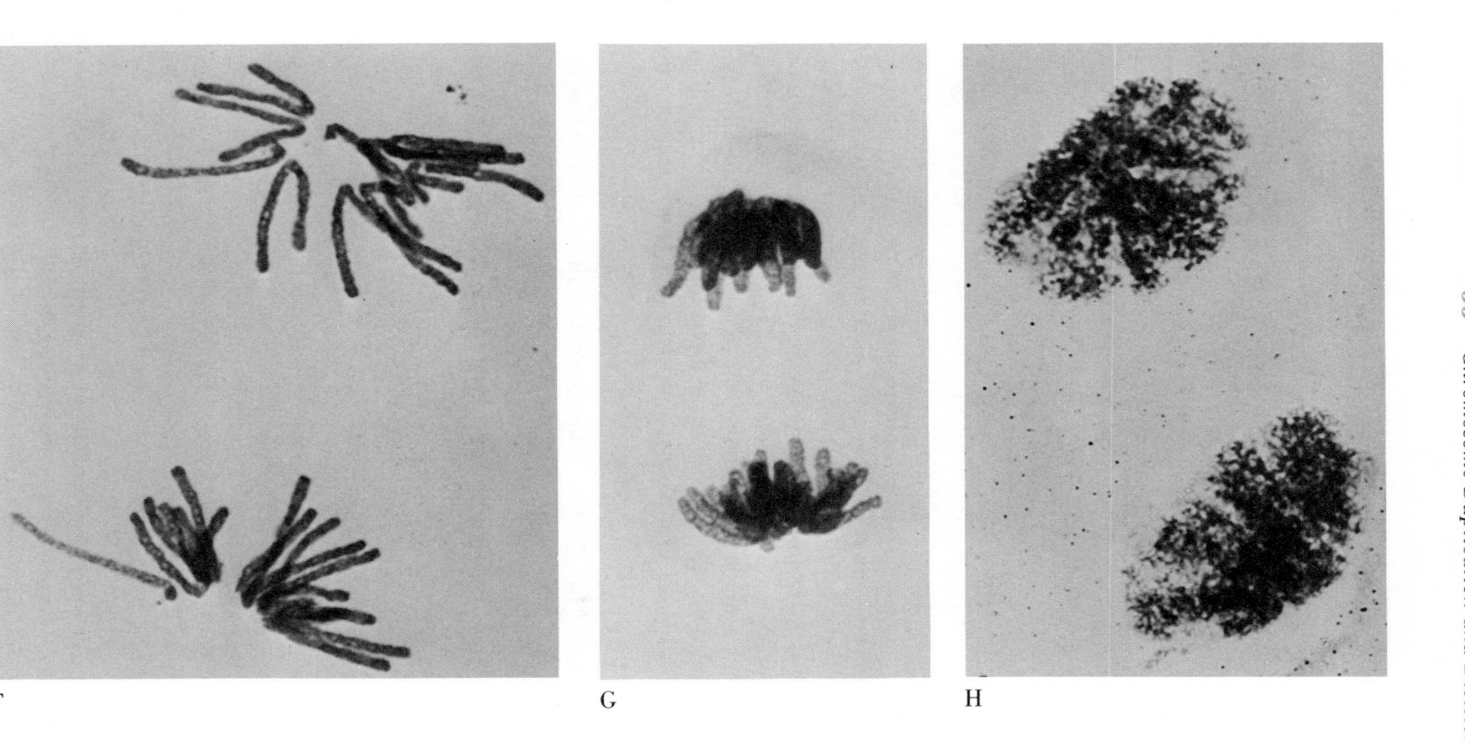

Fig. 2-2 *Photomicrographs of mitosis in the peony. (A) Interphase nucleus, (B) early prophase, (C) late prophase, (D) metaphase, (E) early anaphase, (F) late anaphase, (G) telophase, (H) interphase daughter nuclei. (Courtesy of Marta Walters and Spencer Brown.)*

restored. Chromosome number is a constant characteristic for each species of organism; for example, there are 46 chromosomes in man and 20 chromosomes in maize. If gametes arose by mitosis, their fusion to form a zygote would result in a doubling of the chromosome number at each generation. Thus, in man each egg and each sperm has 23 chromosomes, and the fertilized egg has 46.

Each stage of meiosis (Fig. 2-3A–N) is defined on the basis of the appearance and orientation of the chromosomes, just as in mitosis. Prophase, the first stage of meiosis, is characterized by five substages: *leptotene, zygotene, pachytene, diplotene,* and *diakinesis.*

Leptotene (Fig. 2-3B) At the onset of prophase of the first meiotic division, the chromosomes appear as thin, threadlike structures. The chromosome number characteristic of the organism can often be discerned at this time, and it is seen that there are two of each kind of chromosome. The two members of a pair of chromosomes, or pair of *homologues,* are morphologically identical with respect to size, the position of the centromere, and other characteristics. In maize, for example, there are 10 distinct homologous pairs of chromosomes. As in mitosis, each of the homologous chromosomes is duplicated by early prophase. Thus, each homologous chromosome consists of a pair of sister chromatids.

Zygotene (Fig. 2-3C) The stage of zygotene starts when the homologous chromosomes begin to pair, or *synapse.* Here one of the major differences between meiosis and mitosis can be seen, for it will be recalled that in mitosis there is *no pairing* between homologous chromosomes. The pairing at zygotene appears to be zipperlike and is point for point along the length of the chromosome. The nature of the pairing mechanism is as yet unknown.

Pachytene (Fig. 2-3D) Pachytene is the stage at which pairing is complete, and the chromosomes appear to become thicker and more tightly coiled. The paired homologous chromosomes or *bivalents,* as they are now called, can be seen to be tightly coiled around each other in some organisms.

Diplotene (Fig. 2-3E) At this stage of prophase the bivalent opens longitudinally, and its four chromatids are visible. The centromeres, however, are not split at this time. Furthermore, longitudinal separation is incomplete, and it can be seen that the bivalent is held together at various points along its length. These points of contact are known as *chiasmata* (singular *chiasma*). There may be from one to several of these chiasmata, depending upon the length of the bivalent, and each chiasma may correspond to a point at which two *nonsister* chromatids of a bivalent have undergone an exchange of parts. The genetic implications of such an exchange, known as a *crossover*, will be discussed in detail later.

Diakinesis (Fig. 2-3F) A shortening of the chromosomes that is seen at the diplotene stage of meiosis reaches its extreme at diakinesis, where the chromosomes appear to be tightly coiled.

Metaphase I (Fig. 2-3G and H) Metaphase of the first meiotic division begins when the double nuclear membrane disperses and the spindle forms. An essential difference between meiotic and mitotic metaphase appears at this point. At the first meiotic metaphase each bivalent possesses two independent and undivided centromeres which are arranged above and below the equator and not parallel to it as in mitosis. Furthermore, it can be shown that the different pairs of homologous chromosomes align themselves on the equator of the cell in an independent fashion. Compare, for example, the two possible metaphase alignments of the two pairs of nonhomologous chromosomes shown in Fig. 2-3G and H. These alignments occur with equal frequency and indicate that nonhomologous chromosomes are independent of each other in their arrangement during metaphase I. The significance of this *independent arrangement* will be discussed in Chapter 3. For the remaining stages of meiosis we will follow the results of the metaphase alignment shown in Fig. 2-3G. At the conclusion of the description, work out for yourself the results obtained from the alignment shown in Fig. 2-3H.

Anaphase I (Fig. 2-3I) The first anaphase stage begins when the undivided centromeres and consequently the nonsister chromatids of each bivalent separate and move to opposite poles of the cell. The bivalent is no longer held together by chiasmata, and the sister chromatids are now held together only by their centromeres.

Telophase and Interphase (Fig. 2-3J) After anaphase *I* a double nuclear membrane may be re-formed around the chromosomes, as with the grasshopper; or, as in the plant *Trillium*, no membrane is formed and the chromosomes enter directly into the second meiotic division.

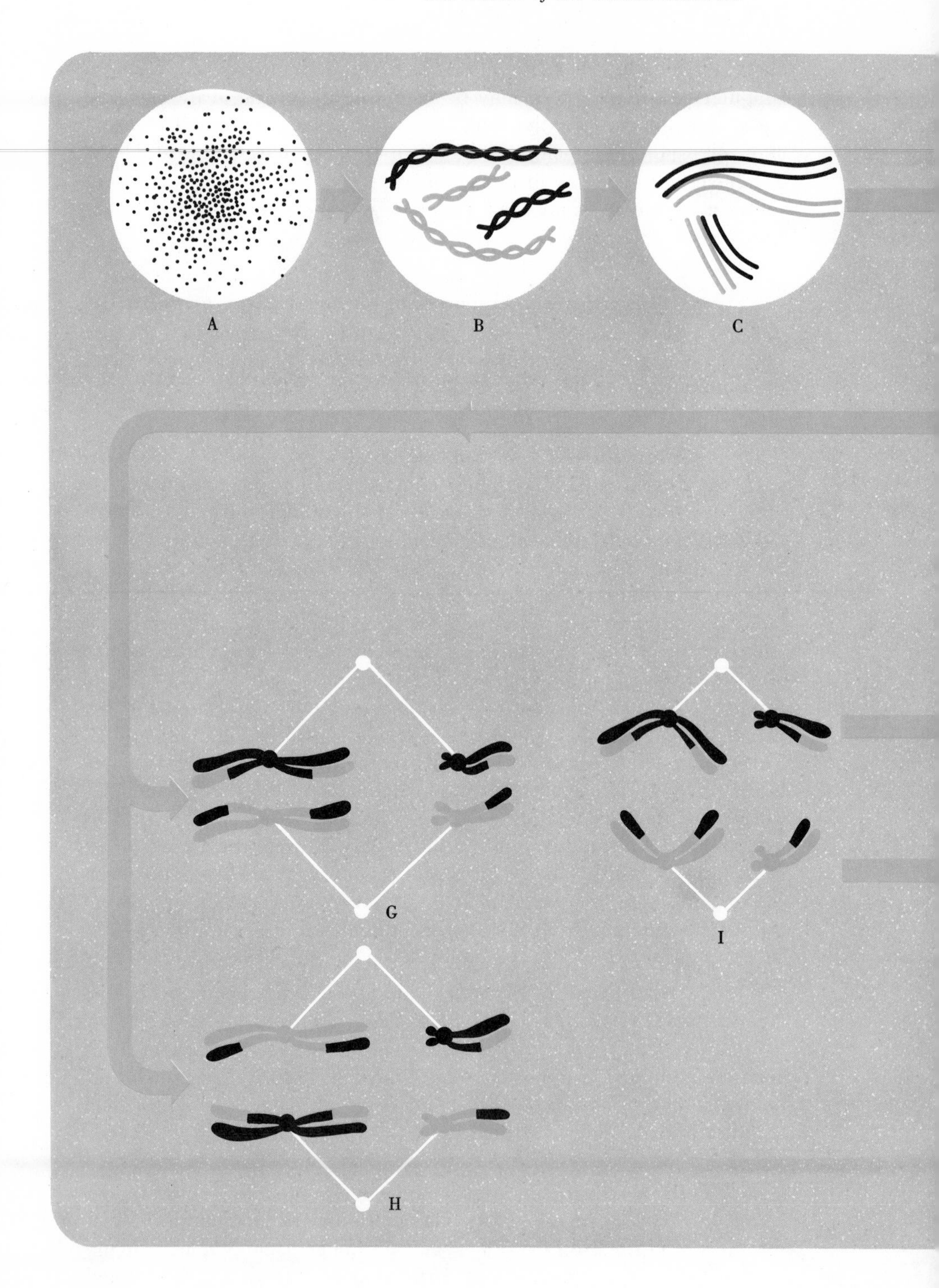
A
B
C
G
I
H

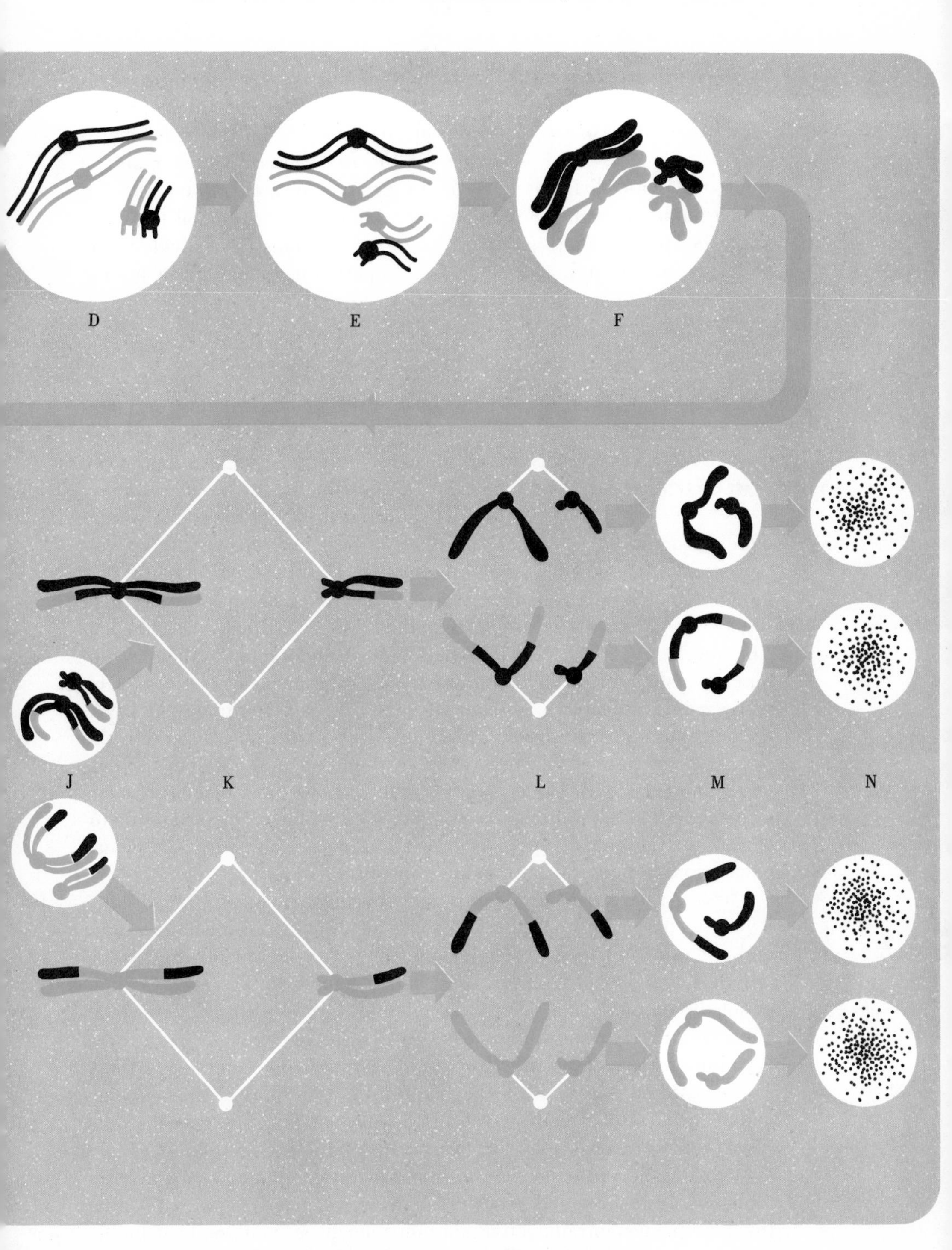

Fig. 2-3 Diagram illustrating the phases of meiosis in an organism with two pairs of chromosomes (see text for details).

Metaphase II and Anaphase II (Fig. 2-3K and L) The second metaphase and anaphase stages are essentially like those of mitosis, and it is *only* at anaphase *II* that the centromeres divide for the first time. The net result of the two meiotic divisions is to yield four cells, or a *tetrad*, each cell of which possesses half the number of chromosomes of the cell that originally began meiosis (Fig. 2-3M). The chromosomes have divided only once, at anaphase I, and their final separation, essentially a mitotic one, has taken place at anaphase II.

MITOTIC AND MEIOTIC DUPLICATION OF CHROMOSOMES AND CHROMOSOMAL DNA

Cytological observations of mitotic and meiotic prophase chromosomes have shown them to have undergone duplication by the time they are visible under the microscope. Thus, it is suggested that the actual duplication process occurs before prophase. Since chromosomal DNA can be detected by specific stains or by the use of radioactive isotopes, it is possible to follow chromosome duplication by following the chromosomal DNA. Photometric determinations, such as those described in Chapter 1, reveal that the DNA has indeed undergone duplication before or just at the onset of prophase. Similar measurements can be made using radioactive isotopes of some of the constituents of DNA. By determining the time at which an isotope is incorporated into the DNA, an estimate of the time of DNA duplication is obtained. In certain root-tip cells, the incorporation of an isotope can be seen to occur some eight hours before the onset of mitotic prophase. During spermatogenesis in the grasshopper, incorporation of an isotope into DNA has been found to occur at premeiotic interphase.

As a result of experiments done by the cytologist J. H. Taylor, it has been possible to suggest a mode of duplication for both mitotic and meiotic DNA in the chromosomes of higher organisms that closely parallels that shown for DNA in *E. coli* (Chapter 1). The mitotic duplication of chromosomal DNA was studied with seedlings of the bean *Vicia faba*. *Vicia* seedlings were grown in solutions of radioactive thymidine, a molecule that contains one of the pyrimidines of DNA. The thymidine was found to be taken up by the roots of the seedlings and incorporated into the DNA of the chromosomes of root-tip cells. Seedlings were grown in the presence of radioactive thymidine long enough for the chromosomes to undergo duplication, after which the roots were washed and transferred to solutions without radioactive thymidine. The nonradioactive solutions, however, contained colchicine, a drug that blocks spindle formation and thus nuclear division, but does not block chromosome duplication. Hence, the distribution of the radioactive label could be followed through more than one chromosome generation within the same cell.

When the chromosomes were examined after duplication in the radioactive thymidine, it was found that each pair of daughter chromosomes was labeled with the isotope. After a second duplication, but in the absence of the isotope, it was found that among the daughter chromosomes one was labeled with the isotope and the other was not. In other words, a labeled chromosome gave rise to one labeled and one unlabeled chromosome. The interpretation of these results is shown in Fig. 2-4. It is based upon the assumption that the chromosome contains at least two DNA units. In the formation of daughter chromosomes these units separate and each unit duplicates. When this duplication occurs in the presence of radioactive thymidine, the two daughter chromosomes will be labeled. A second duplication in the absence of the isotope will thus lead to one labeled and one unlabeled chromosome. This result is in accord with the structure of DNA and its mode of duplication as proposed by Watson and Crick (Fig. 1-13).

Labeled thymidine was also used by Taylor to investigate the meiotic duplication of chromosomes in males of the grasshopper,

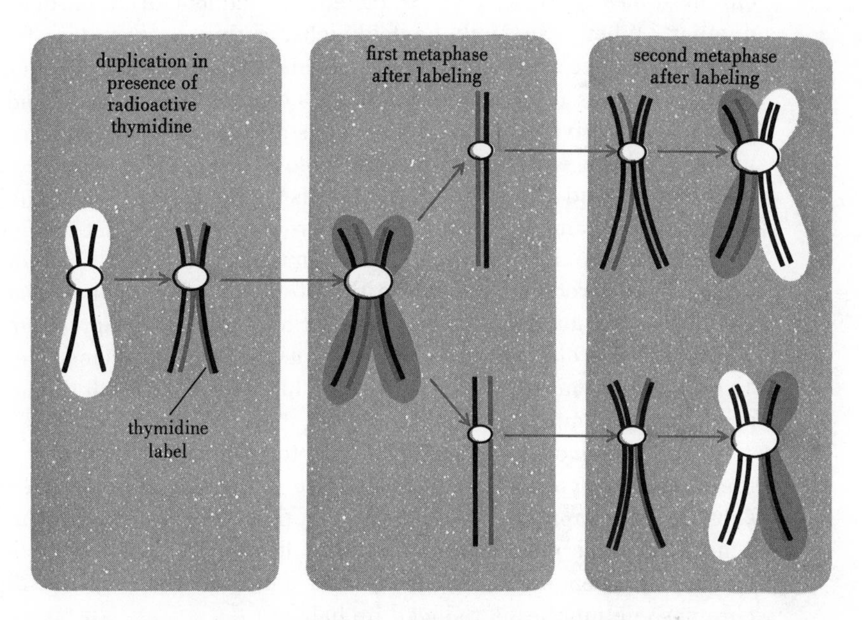

Fig. 2-4 *The mitotic duplication of chromosomes; the chromosomes labeled with radioactive thymidine are shown in color (see text for details). (After Taylor,* American Naturalist, *Vol. 91, 1957.)*

Romalea microptera. The isotope was injected into the animals and testes were removed from different animals at various time intervals. The distribution of the label in the meiotic chromosomes during spermatogenesis was found to be similar to that found for mitosis. In other words, the DNA of the meiotic chromosomes also duplicates in a semiconservative fashion. Of significance is the fact that Taylor's observations, obtained with intact chromosomes, are in accord with the observations derived from density-gradient centrifugation described in chapter 1.

LIFE CYCLES

The major types of animal, microbial, and plant life cycles are discussed in *Animal Structure and Function*, *Microbial Life*, and *Plant Diversification*, other volumes in this series. The life cycles of all sexually reproducing organisms are characterized by having two phases. One of these phases, the *haploid* phase, consists of cells that have been *derived by meiosis;* the other, the *diploid phase*, consists of cells, some of which are capable of *undergoing meiosis*.

Meiosis yields haploid cells, eggs and sperm for example, in which the chromosome number has been reduced by half. The number of chromosomes in these cells is called the haploid or N chromosome number. When two such haploid cells, acting as gametes, fuse at fertilization, a cell termed a *zygote* is formed. Each kind of chromosome in the zygote is represented twice. Its chromosome number, and that of any cells derived from it by mitosis, is the diploid or $2N$ chromosome number. These cells represent the diploid phase of the life cycle. In man, the haploid chromosome number is 23. When fertilization occurs, each haploid gamete contributes 23 chromosomes, so that a zygote is formed with the diploid chromosome number of 46. Clearly, the diploid phase is the predominant one in the life cycle of man. It is also true for most other animals. Among plants the diploid phase predominates in the life cycle of most multicellular forms. On the other hand, the fungi as well as the unicellular algae possess life cycles in which the haploid phase is predominant.

Three life cycles have been selected for description here since they relate to types of organisms whose genetics will be discussed in subsequent chapters. Although many other sexually reproducing organisms exhibit modifications of these life cycles, none differ in the underlying phenomenon of meiosis as a means whereby the diploid chromosome number is reduced by half.

The three life cycles chosen for description here are those of a unicellular green alga, *Chlamydomonas reinhardi*; a fungus, *Neurospora crassa*; and a fruit fly, *Drosophila melanogaster*. Hereafter, for simplicity, these three organisms will be referred to by their generic names.

There are two sexes of *Chlamydomonas*, referred to as mating type *plus* and *minus* (mt^+ and mt^-). Although the two sexes are identical in morphology and size, they are physiologically distinct, for when they come into contact with each other they can act as gametes and fuse to form a diploid zygote (Fig. 2-5). The zygote does not divide mitotically; rather it undergoes meiosis to yield a *tetrad* of four haploid meiotic products, two that are mt^+ and two mt^-. Each of these meiotic products then divides mitotically to produce clones of genotypically identical haploid vegetative cells. Under the appropriate conditions, any one of these vegetative cells can become differentiated into a gamete, and the life cycle is thus repeated when gametes of the opposite mating type fuse. The predominant phase of the life cycle in *Chlamydomonas* is haploid.

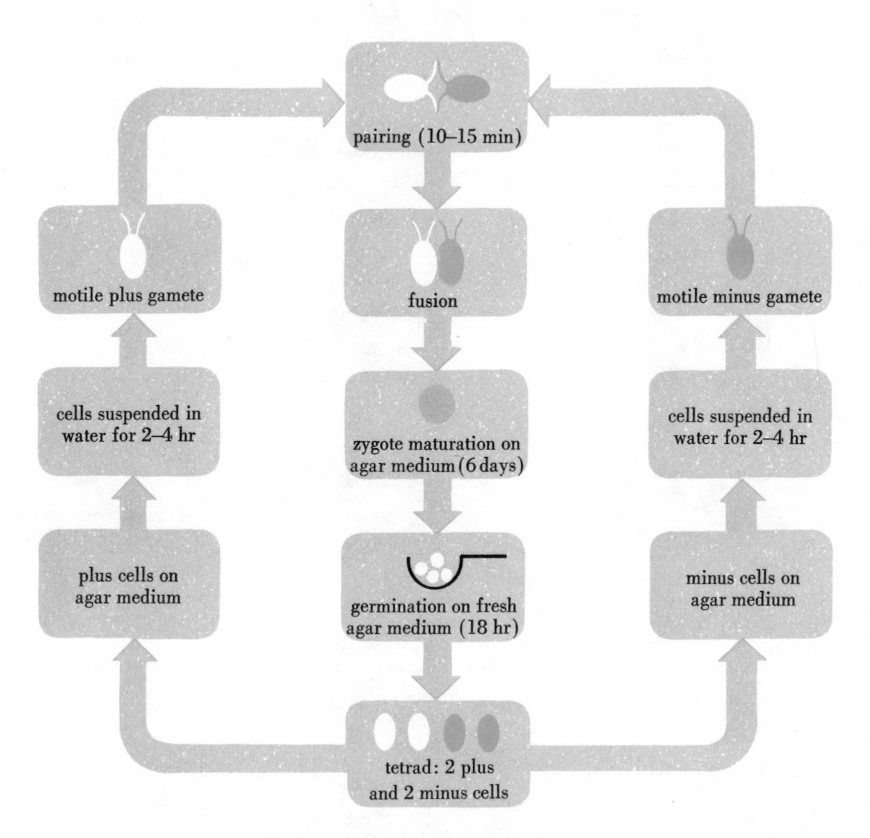

Fig. 2-5 *Life cycle of* Chlamydomonas *(description in text). (Redrawn from Levine and Ebersold,* Cold Spring Harbor Symposia on Quantitative Biology, *Vol. 23, 1958.)*

Neurospora also has a life cycle that is predominantly haploid, and the two haploid mating types (*A* and *a*) can reproduce vegetatively by mitosis (Fig. 2-6). Repeated mitoses produce long filaments called *hyphae*. Certain cells of the hyphae can form spores called *conidia* (singular *conidium*). Conidia can either give rise to a new colony of

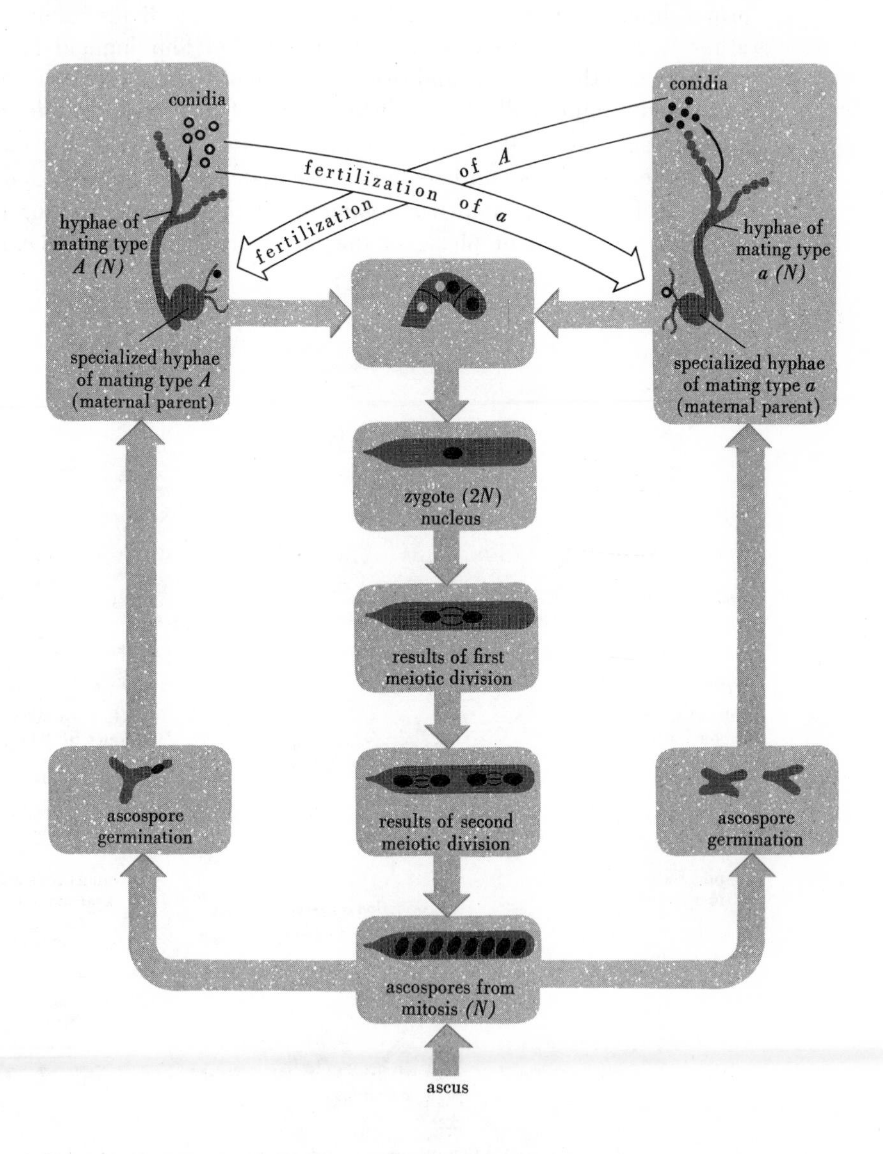

Fig. 2-6 *Life cycle of* Neurospora *(description in text). (Redrawn from Wagner and Mitchell,* Genetics and Metabolism. *New York: Wiley, 1955.)*

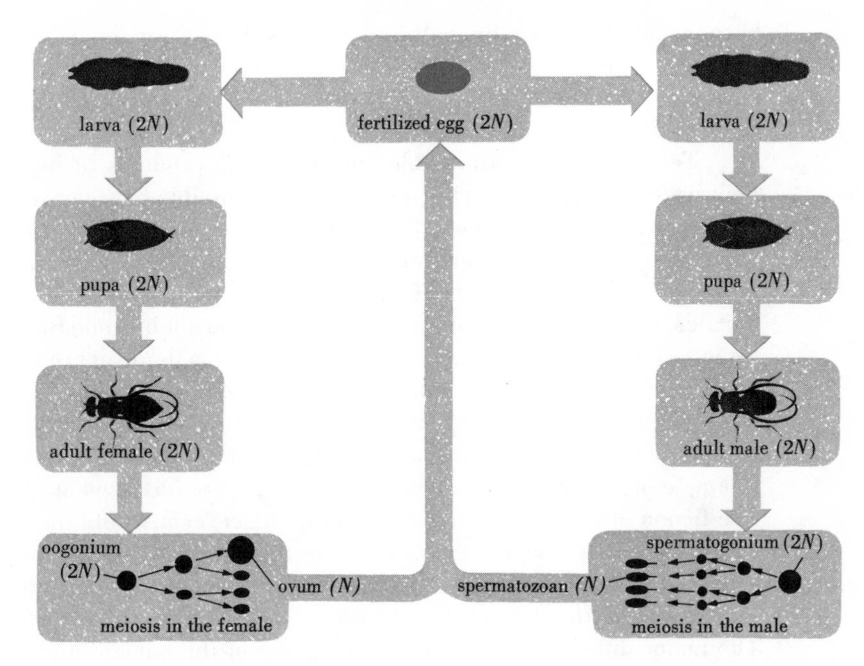

Fig. 2-7 Life cycle of Drosophila *(description in text).*

vegetative hyphae, or take part in sexual reproduction. Under appropriate conditions either mating type can form a specialized hyphal cell which will fuse with a conidium from the opposite mating type. Two nuclei are now present within a single cell. These nuclei undergo several mitotic divisions. Eventually, two nuclei derived from cells of opposite mating type will fuse, forming a diploid nucleus. The diploid nucleus is now located in a cell called an *ascus* (plural *asci*). Meiosis within the ascus will yield four haploid meiotic products or four haploid nuclei. Each of these nuclei, in turn, goes through a *mitotic division.* Thus, each meiotic product is represented *twice* in the ascus. The ascus, therefore, contains eight haploid nuclei, and a spore wall forms around each nucleus, forming eight *ascospores*. The ascospores are released when the ascus ruptures, and after each ascospore germinates it will give rise to haploid hyphal cells and haploid conidia through mitotic division.

Both *Neurospora* and *Chlamydomonas* possess an important feature for genetic research. Asci of *Neurospora* and the mature zygotes of *Chlamydomas* can be isolated individually, and each of the four meiotic products can be easily separated from either the ascus or the zygote wall. In other words, it is possible to obtain and separate the four products from a single meiosis. After each of these products has been isolated, it can be cultured and its hereditary characteristics analyzed. The analysis of all meiotic products from many *individual* meioses is called *tetrad analysis*. It will be discussed in detail in Chapter 3.

Tetrad analysis is possible only when the products of individual meioses can be isolated. This isolation is not possible in organisms having a predominantly diploid life cycle. In the case of *Drosophila*, or other animals for that matter, the haploid phase, which arises by meiosis, is very brief and difficult to isolate. Also, as in the case of females, three out of the four meiotic products do not become functional gametes. As shown in Fig. 2-7, meiosis, in the male, leads to the production of four meiotic products or *spermatozoa*. These, however, mix freely with other spermatozoa. In the female, meiosis leads to the production of only one *ovum* from an *oogonial* cell. The female fly receives a sample of sperm from the male, and the eggs are fertilized at random. The fusion of the haploid egg and sperm nuclei reconstitute the diploid stage. In the case of *Drosophila*, the fertilized egg will hatch and two stages—the larval and pupal—precede the formation of an adult fly.

Each of the life cycles described here is basically the same. The major difference lies only in the extent of the haploid and diploid phases, and in the fact that organisms with a haploid life cycle present an opportunity for the geneticist to obtain for analysis all of the haploid products from individual meioses.

FURTHER READING

Mazia, D., 1961. "Mitosis and the physiology of cell division," in *The cell*, vol. 3, *Meiosis and mitosis*, pp. 77–412, Brachet, J. and Mirsky, A. E. (eds.). New York: Academic Press Inc.

Rhoades, M. M., 1961. "Meiosis," in *The cell*, vol. 3, *Meiosis and mitosis*, pp. 1–75. Brachet, J. and Mirsky, A. E. (eds.). New York: Academic Press Inc.

Swanson C., 1957. *Cytology and cytogenetics*. Englewood Cliffs, N.J.: Prentice-Hall, Inc.

part II

THE TRANSMISSION OF THE GENETIC MATERIAL

chapter 3

The Segregation of Genes

Several of the phenomena observed at meiosis are of importance in understanding how the genetic material of nucleate organisms is transferred from one generation to the next. As described in Chapter 2, homologous chromosomes pair at zygotene and separate at anaphase I. Thus, if we were to label one member of a chromosome pair as *A* and the other as *a*, and follow them through the completion of meiosis, we would see that of the four meiotic products or tetrad, two would have *A* and two would have *a*. As Fig. 3-1 shows, this is a consequence of the pairing of homologous chromosomes at zygotene and their separation at anaphase I. A question arises, therefore, as to the relation between this behavior of the chromosomes at meiosis and the transmission of a genetic trait.

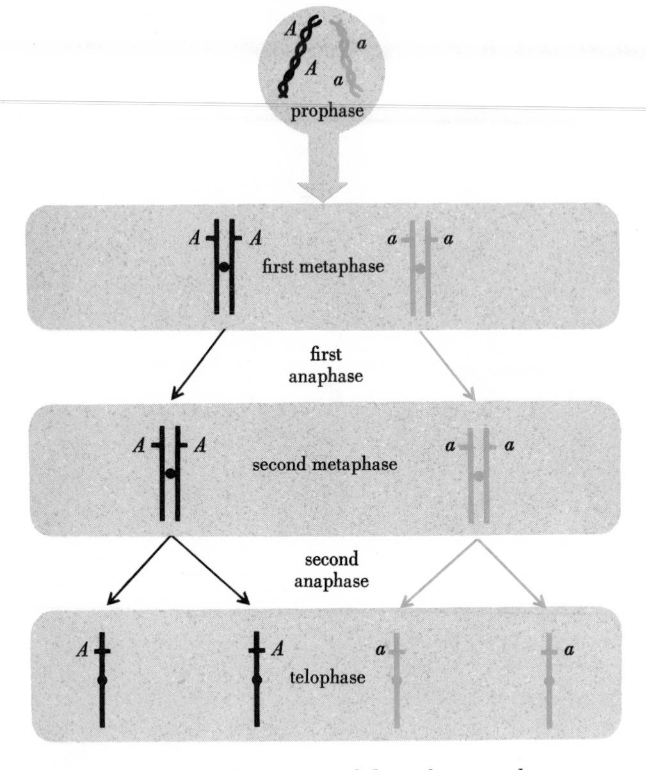

Fig. 3-1 The separation of a pair of homologous chromosomes carrying their alleles A *and* a *during meiosis.*

In order to answer this question, a hypothesis can be formulated, relating the transmission of a given genetic trait to the behavior of a pair of homologous chromosomes at meiosis. First, the genetic material, DNA, is found within the chromosomes, and it is organized into units called *genes*. A gene has the genetic information necessary for the determination of a specific genetic trait. Second, assume that such a gene, denoted as *A*, occupies a particular site or *locus* in a given chromosome. Further, the homologous chromosome contains at the identical locus an alternative gene *a*, which controls the same trait as *A* but in such a way as to result in the production of a different phenotype for that trait. These alternative genes at the same locus, *A* and *a*, are called *alleles* (from the Greek word *allelon*, meaning of one another). Our hypothesis can now be stated in the following way: *If one homologue contains the gene A and the other homologue its allele, gene a, the two genes—just as the two homologous chromosomes—should be found in a one-to-one ratio among the meiotic products.*

The hypothesis can be tested by controlled breeding experiments or crosses with a variety of animals and plants. The purpose of this chapter is to present some of the experimental evidence that shows the hypothesis to be correct. The facts derived from these experiments form part of a large body of evidence that supports the *chromosomal theory of inheritance.* This theory will be developed here and in the chapters to follow. Before the evidence in favor of the foregoing hypothesis is presented, it is necessary to define certain terms used in genetic experiments.

The study of genetics depends to a very great extent upon crosses or breeding experiments involving individuals having genetic differences. These differences are seen as pairs of alternative phenotypes, such as colored versus albino fur in rabbits or virulent versus avirulent pneumococci. Since most methods of genetic analysis depend upon a comparison of one phenotype with another, there must be, for any organism under investigation, some standard type for comparison upon which all geneticists can agree. It is inevitable that a standard type will be somewhat arbitrary, but no more so than any other standards such as the yard and the meter, or the kilogram and the pound. In genetics, the standard type for any experimental organism, whether it is a bacterial virus, a fruit fly, or a corn plant, is called the *wild type.* The term stems from the idea that this is the sort of organism found in nature—that is, in the wild—rather than in the laboratory or under domestication. We can speak of the wild type as being a certain combination of phenotypic characteristics. Most often, however, we will concern ourselves with *only one* of the many characteristics of the organism, and say that it possesses the wild-type phenotype for that trait, neglecting the other phenotypic traits it may show.

In contrast to the wild-type phenotype, there are departures known as *mutant* phenotypes. They are the result of heritable alterations in the genotype that have arisen by a process known as *mutation.* In Chapter 9 the question of mutation will be examined in detail. Among examples of wild-type and mutant phenotypes, the smooth form of pneumococcus can be considered as being the wild type and the rough form the mutant. Wild-type *Chlamydomonas* is green, but there is also a yellow mutant form. Wild-type *Neurospora* can synthesize all of the amino acids necessary for its growth, but there are many different mutant phenotypes, each characterized by the inability to synthesize a given amino acid. Wild-type *Drosophila* has a red-brown eye color, but there is a mutant phenotype that is white, because of a loss in the mutant organism's ability to synthesize eye pigments.

In order to be concise, as well as unambiguous, geneticists have chosen specific kinds of notation to use when referring to wild-type and mutant alleles. Several systems ot notation have arisen, mainly because geneticists work with different organisms. Throughout this book, a single system of notation will be used where practical. When a given wild-type allele is being considered, the symbol + will be used. The mutant allele, on the other hand, will be symbolized by one, two, or three letters that are an abbreviation of the trait being considered. For example, a strain of *Chlamydomonas* unable to synthesize the amino acid arginine carries the mutant allele symbolized as *arg*, and the wild type carries the + allele.

When the wild-type or mutant strains are propagated by vegetative reproduction, in the case of organisms with a haploid life cycle, or by inbreeding wild type with wild type or mutant with mutant for those with a diploid life cycle, the genetic characteristics of the two strains are retained through countless generations. What happens, however, when a wild-type and a mutant strain are crossed?

SEGREGATION–TETRAD ANALYSIS

We can answer the foregoing question by making a cross with *Chlamydomonas*. Wild-type *Chlamydomonas* is photosynthetic and can synthesize all of the substances necessary for its growth and reproduction when it is grown in the light in the presence of inorganic salts, carbon dioxide, and water. This simple medium is called a *minimal medium*. Mutant strains of *Chlamydomonas*, and other microorganisms unable to synthesize a particular amino acid or vitamin, are called *nutritional* or *biochemical mutants*. They will grow only if the minimal medium is supplemented with the specific compound that the mutant is unable to synthesize.

The wild-type strain carries within one of its chromosomes a wild-type *gene* that controls the synthesis of arginine. A mutant strain carries, in the homologous chromosome, a mutant gene or allele and is unable to synthesize the amino acid arginine. Hence, if the two strains are crossed, the behavior of the pair of homologous chromosomes at meiosis should result in the production of meiotic products half carrying the + gene and half carrying the *arg* gene.

The + and *arg* strains of *Chlamydomonas* can be crossed if they differ in mating type, and a cross of +, $mt^+ \times$ *arg*, mt^- is easily achieved. Several thousand gametes of each parental type are mixed together, and after a few minutes pairs of opposite mating type will fuse to form zygotes. After maturation, several hundred zygotes are transferred to minimal medium supplemented with arginine and separated from each other. Five zygotes have been transferred in the example beginning with Fig. 3-2A. After the zygotes have completed meiosis and germinated

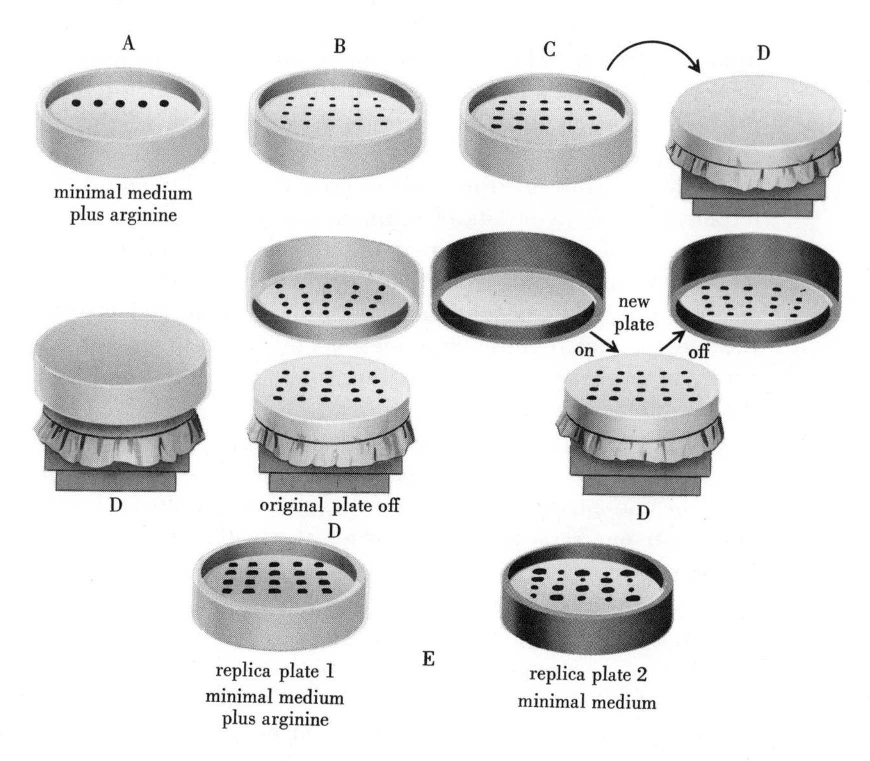

Fig. 3-2 *Tetrad analysis of the cross* arg × + *in* Chlamydomonas.

(Fig. 3-2B), the four meiotic products, or tetrad, of *each zygote* are separated from each other and allowed to reproduce mitotically to form four colonies (Fig. 3-2C).

The genetic nature of these colonies is then ascertained by a process known as *replica plating*, whereby cells of each colony can be transferred from the original petri plate to an absorbent surface such as filter paper or velvet. When fresh culture medium contained in petri plates is pressed onto the surface of the filter paper, a few cells are transferred from it to the surface of the medium (Fig. 3-2D). These cells can grow to form colonies, and the new plate is thus an exact replica of the original. It is possible to make several replicas from the original medium.

In this experiment the colonies on the original medium containing arginine are replica plated to minimal medium, and to minimal medium containing arginine (Fig. 3-2E). After about 72 hours, colonies are visible on both kinds of medium. On the replica plate supplemented with arginine, each of the original tetrads is represented by four colonies. These four colonies represent, in turn, the original meiotic products from each zygote. The plate containing the minimal medium, however, shows that colonies have developed from only two of the four meiotic products of each tetrad. Microscopic examination of this plate will show that cells from the other two meiotic products were transferred by replication, but that they did not grow on the minimal medium. Each tetrad, therefore, consists of *equal numbers of two kinds of cells*. Half of the cells of each tetrad grow on either minimal medium or minimal medium with arginine, and hence are wild type. The other half grow *only* on the medium supplemented with arginine and are, therefore, the mutant that requires arginine for growth. In other words, the two parental types, + and *arg*, are represented in equal numbers among the products of meiosis.

We can obtain the same result by making the *reciprocal cross*, that is, +, mt^- × *arg* mt^+, indicating that the sex or mating type has no effect on the *one-to-one* ratio of +:*arg* meiotic products. Furthermore, this one-to-one ratio is obtained when the wild-type strain of *Chlamydomonas* is crossed with any one of several different mutant strains. Analysis of tetrads from similar crosses with *Neurospora* gives one-to-one ratios of wild type to mutant.

Let us now examine the cross between the + and *arg* strains of *Chlamydomonas* in terms of the events occurring at meiosis. If we assume that the + and *arg* genes are in homologous chromosomes, it is evident that the behavior of the pair of genes or alleles can be explained on the basis of the behavior of the pair of homologous chromosomes during meiosis (Fig. 3-1). We see that the homologous chromosomes pair at zygotene, thus bringing together the two alleles, + and *arg*. At some point in meiosis, probably preprophase, the homologues have undergone duplication, so that at zygotene each of the alleles is represented twice. At anaphase I, we note that the alleles now *segregate* from each other at the time of *separation* of nonsister chromatids. When the sister chromatids separate, the second meiotic division results in two meiotic products with the *arg* allele and two with the + allele.

The hypothesis formulated on page 48 and the experimental results just discussed are in agreement. There is a correlation between the one-to-one *separation* of a pair of homologous chromosomes and the one-to-one *segregation* of a pair of genes or alleles. The phenomenon of segregation is referred to as the Principle of Segregation. This principle or law was one of several formulated in 1866 by an Austrian monk, Gregor Mendel, on the basis of experiments concerned with inheritance in the garden pea. This principle is basic to the chromosome theory of inheritance, and it has been shown to hold for the inheritance of a pair of alleles, wherever tested.

SEGREGATION—STRAND ANALYSIS

Tetrad analysis is not possible in higher animals and plants whose life cycle is predominantly diploid. If you will recall the life cycle of *Drosophila*, for example, you will note that in oogenesis only one of the four meiotic products functions as the egg, and the choice as to which will function as the egg is random. Moreover, this egg is fertilized at random by one of many sperm. Thus, any genetic analysis of a diploid organism is based upon a random sampling of meiotic products rather than a sampling of all of the products of individual meioses. An analysis of this sort is called *strand analysis*. The term strand refers to any one chromosome in a meiotic product.

Such different diploid organisms as corn, mice, and the fruit fly *Drosophila* are used in genetic studies. *Drosophila* has been used for genetic investigations since shortly after the turn of the century, and there are many different mutant strains. Most of these mutant strains are of a morphological sort, differing from wild type with respect to the size or shape of wings and eyes, the color of eyes and body, or the number, pattern, and shape of bristles. When any mutant strain is inbred it reproduces true to type, and mutant strains have been maintained as laboratory cultures by inbreeding for several hundred generations. One of the well-known mutant strains of *Drosophila*, *vestigial (vg)*, has extremely short, stubby and abnormally shaped wings. It is to be contrasted with the wild-type strain in which the pair of wings extends over the abdomen (Fig. 3-3).

When + and *vg* flies are crossed, regardless of which is the male or female parent, all of the progeny are phenotypically wild type; none of the flies show the *vestigial* phenotype. Flies of this F_1, or *first filial generation* following the cross of + and *vg*, can be crossed again to wild-type flies, and once more all of the progeny show the wild phenotype. However, if they are crossed to *vestigial* flies, a different result is obtained. The results of a typical experiment in which the F_1 flies are crossed with *vestigial* flies are shown in Fig. 3-3. Out of 984 progeny from the cross of the F_1 flies to *vestigial* flies, approximately one half are wild type and one half are *vestigial*. The results of this cross can be explained by the hypothesis put forth in the introduction to this chapter. That is, there is a pair of genes, + and *vg*, that control wing development; these genes segregate from each other at meiosis when the two homologous chromosomes that bear them separate at anaphase I. However, *Drosophila* is diploid, unlike *Chlamydomonas* or *Neurospora*, and therefore the parents and their offspring carry a *pair* of these genes. The wild-type strain used in the original cross must have the genotype +/+, and that of the *vestigial* strain must be *vg/vg*. Meiosis in the former will produce gametes *all of which* carry

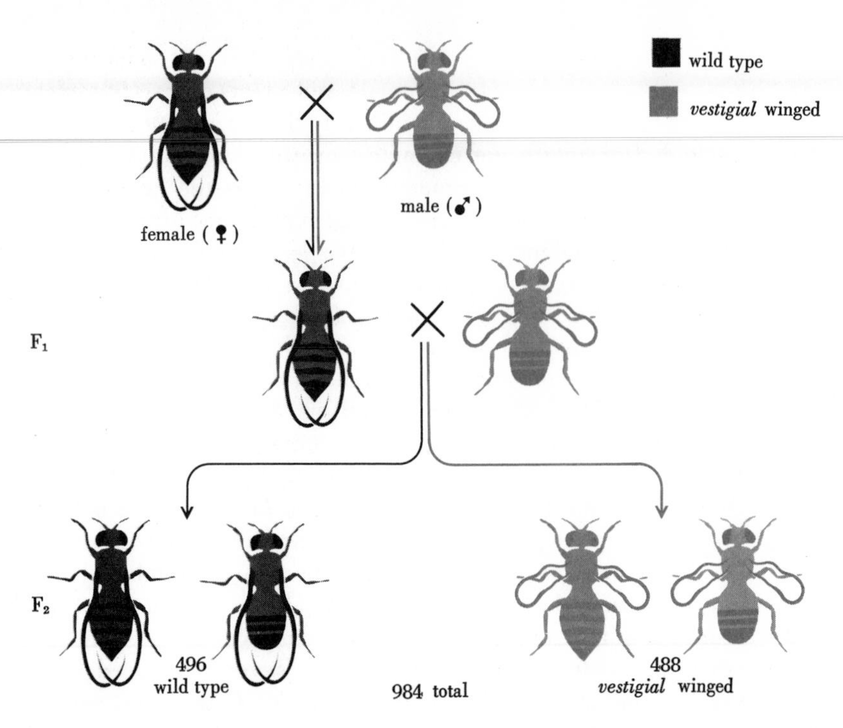

Fig. 3-3 Phenotypes resulting from a cross in Drosophila *between wild-type and* vestigial *flies.*

the + allele, whereas in the latter *all gametes* carry the *vg* allele. The F_1, though *phenotypically* wild type, must have the genotype +/*vg* since it results from the fusion of + and *vg* gametes. Meiosis in the F_1 flies, according to the hypothesis, should yield two kinds of meiotic products, + and *vg*, in equal numbers. If this statement is true, then a cross of the F_1 +/*vg* flies to *vg*/*vg* flies would give ideally 50 percent +/*vg* flies and 50 percent *vg*/*vg* flies. This is shown diagrammatically in Fig. 3-4, along with the numerical data from the previous figure. A one-to-one ratio is expected, and the observed values do not differ from it significantly. Figure 3-5 shows these same results in terms of meiosis.

This cross of +/*vg* flies with *vg*/*vg* flies is called a *test cross* because it *reveals the genotype* of the F_1 flies. It shows that half of the gametes produced by the F_1 +/*vg* flies carry the *vg* allele and half carry the + allele. Even though the F_1 flies are wild-type in phenotype, they have the *vg* gene. The wild phenotype results from the fact that the + gene masks the effect of the *vg* gene. This masking effect is called *dominance*, and it is one of the important genetic phenomena described by Mendel in his original work with peas. The development of a normal wing comes about in the presence of *at least one + gene*, whereas the

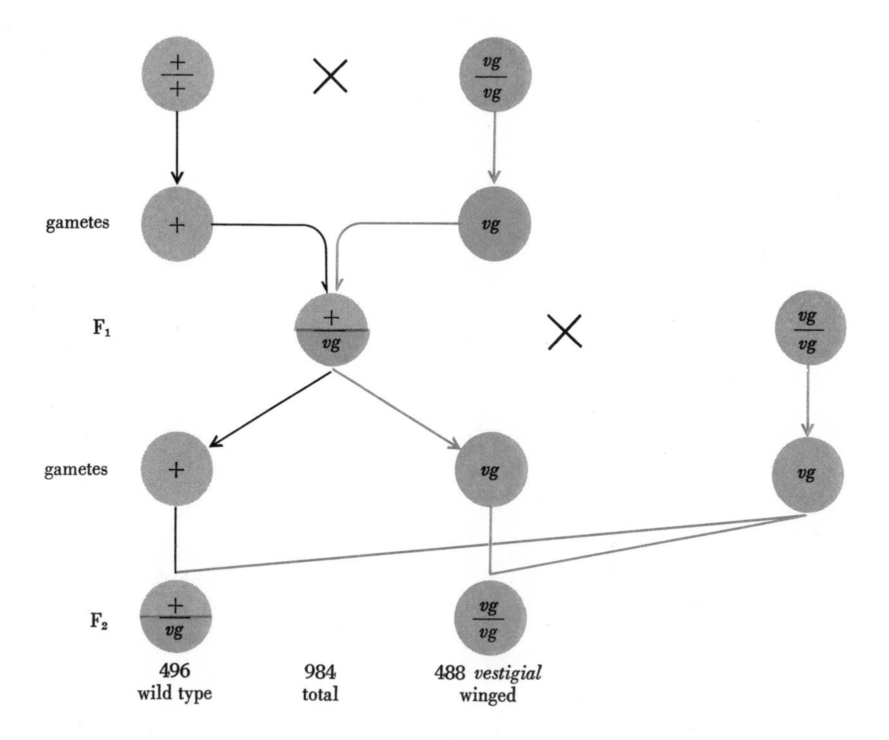

Fig. 3-4 The cross shown in Fig. 3-3 given in terms of genotypes.

development of a *vestigial* wing requires that the *two vg genes* be present. The + gene is *dominant* and the *vg* gene is *recessive* to it.

Two more terms must be introduced before we proceed further. The *vestigial* flies produce only one kind of gamete with respect to the wing phenotype. Such individuals are spoken of as being *homozygous (vg/vg)*. They result from the union of two *vg* gametes and consequently produce only *vg* gametes. The wild-type flies may be either homozygous or *heterozygous*. The original wild-type strain was defined as breeding true to type, and, therefore, each of the flies of this strain *arose* from the union of two + gametes and they *produce* only + gametes. Hence, they are homozygous (+/+) for the + allele of *vg*. However, the F_1 wild-type flies arise from the union of a + and a *vg* gamete. Therefore, they are heterozygous (+/*vg*), but show the wild phenotype because of dominance. A test cross is one way to reveal the fact that they are heterozygous, since they are crossed with homozygous recessive *vg* flies. As a general rule, a test cross is used to reveal the presence of a heterozygous genotype when there is dominance. In the present case the test cross shows that the F_1 flies are indeed +/*vg* and, what is more, that the + and *vg* gametes are produced in the one-to-one ratio expected

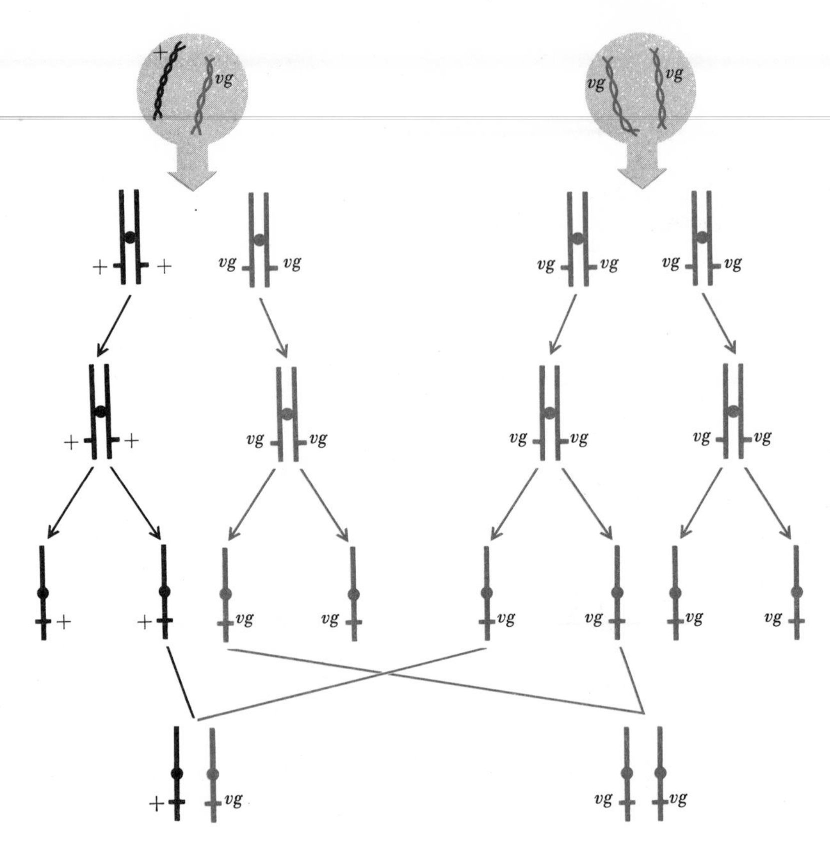

Fig. 3-5 *The results of the cross shown in Fig. 3-3 interpreted in terms of meiosis.*

by our original hypothesis. It can be said, therefore, that the *vg* and + alleles obey the Principle of Segregation.

What occurs, however, if the F_1 flies are intercrossed instead of test crossed? The test cross has shown that the + and *vg* alleles segregate in a one-to-one fashion. Since + and *vg* gametes are produced in a one-to-one ratio, inbreeding the F_1 should give an F_2 that arises from random combinations of egg and sperm that bear either of the two alleles. In other words, a gamete bearing the + allele from a male should have an equal probability of fusing with either a + or *vg* bearing gamete of a female, and vice versa. The possible unions of gametes are four: namely, + with +, + with *vg*, *vg* with +, and *vg* with *vg* (Fig. 3-6). Thus, with respect to the *ratio of genotypes*, approximately one quarter of the flies are +/+, one half are +/*vg*, and one quarter are *vg*/*vg*. Since the + gene is dominant to the *vg* gene, the *ratio of phenotypes* will be three fourths wild type to one fourth *vestigial*. The results of inbreeding the F_1 heterozygotes are given in Fig. 3-6, along with the numerical data. It is seen that the data do agree with the predicted phenotypic ratio of three fourths wild type to one fourth *vestigial* ($3 + : 1$ *vg*). Once again the Principle of Segregation applies, since the results of inbreeding the F_1 are exactly as predicted from the behavior of the chromosomes at meiosis. Work this out for yourself using diagrams similar to those in Fig. 3-5.

One additional test can be made with the flies in this cross. The wild-type flies in the F_2 are either homozygous or heterozygous, and from the expected genotypic ratio, one third of these should be +/+, and two thirds +/*vg*. This can be verified by test crossing individual F_2 wild-type flies to *vg*/*vg* flies.

The Principle of Segregation is exhibited in the inheritance of human traits. A classical example from human inheritance is the case of the skeletal abnormality called *brachydactyly*, in which the fingers are very short. A pedigree for brachydactyly is shown in Fig. 3-7.

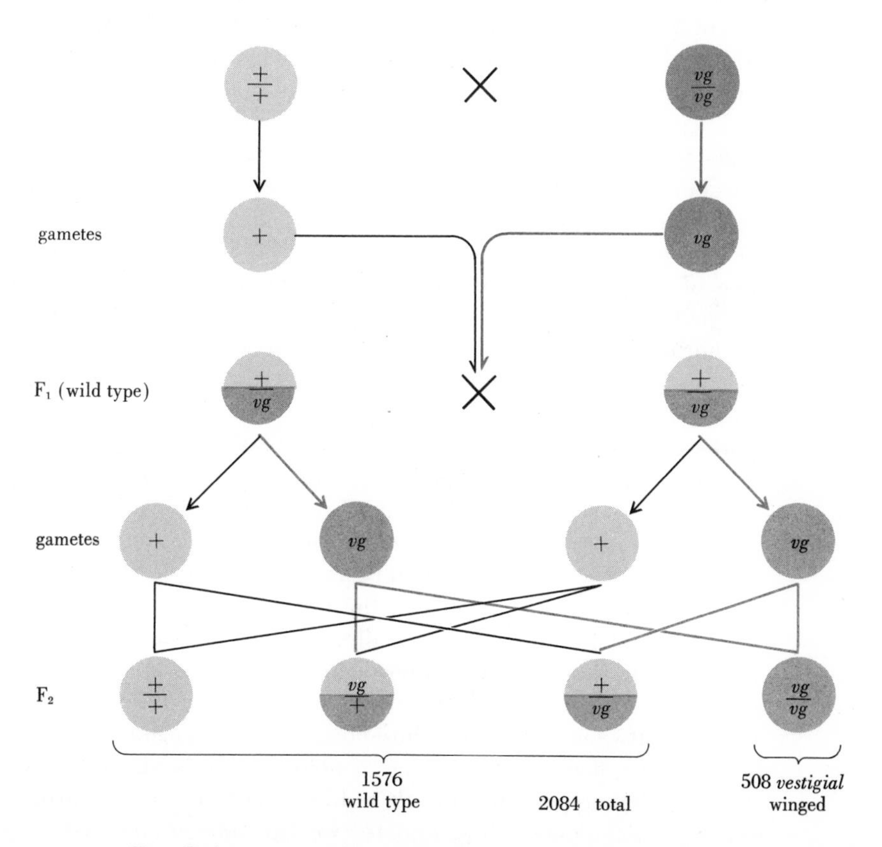

Fig. 3-6 Results from inbreeding +/vg flies in Drosophila.

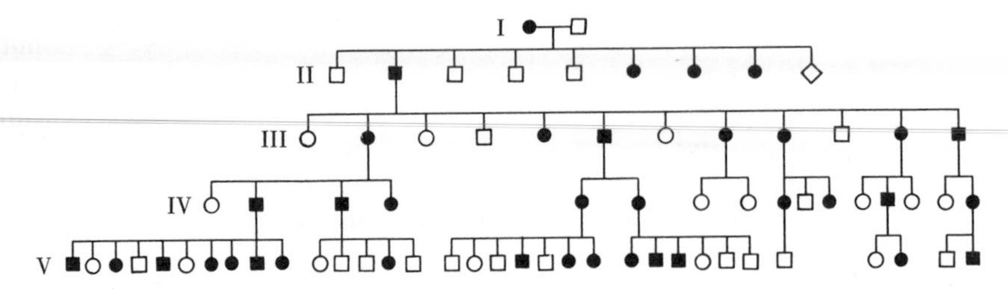

Fig. 3-7 *Pedigree of brachydactyly in man. (After Farabee,* Papers of the Peabody Museum, *Harvard University, Vol. 3, 1905.)*

It was obtained in 1905 and represents one of the first studies of single-gene inheritance in man. The females in the pedigree are represented by circles, and the males by squares. When the sex of the individual is unknown, a diamond-shaped symbol is used. A marriage is shown by a horizontal line connecting a male and a female. A black circle or square represents an individual expressing the trait. Brachydactyly *(B)* is a dominant trait, and, therefore, it is expressed in both heterozygous and homozygous individuals. The normal condition or wild type (+) is recessive to it. If an affected individual is homozygous, then all of his offspring will also be affected. If one parent is heterozygous and the spouse is normal, then on the average, half of the offspring will be affected and half will be normal. The marriages shown in Fig. 3-7 are those of affected individuals to normal individuals, and for the sake of simplicity only the initial marriage at generation I is shown. The pedigree shows that at each generation about half of the individuals are normal and half are affected. When the symbols *B*/+ for heterozygous brachydactylous individuals and +/+ for normal individuals are substituted, you will see that the marriages at each generation in reality represent test crosses. The phenotypes of the children appear in a one-to-one ratio, as predicted by the Principle of Segregation.

We have seen examples of the Principle of Segregation in *Drosophila* and man. Both deal with pairs of alleles in which one is completely dominant to the other. Dominance is attributable to the presence of an allele whose action results in the production of a given phenotype, while the action of the recessive allele may be missing or in some way altered or masked. There are, however, examples of incomplete dominance in which one allele does not entirely mask the other. For instance, a cross between red- and white-flowered snapdragons results in an F_1 with pink flowers. When the F_1 plants are inbred, the F_2 consists of three phenotypes: one fourth red, one half pink, one fourth white. Because dominance is incomplete, we can immediately assign genotypes to each of the three phenotypes; that is, the pink-flowered plants are heterozygous (+/*r*), while the red- and white-flowered plants are homozygous (+/+ and *r*/*r* respectively). Thus, regardless of dominance, the Principle of Segregation applies.

All of the crosses described here demonstrate the Principle of Segregation, and other examples drawn from *Drosophila*, mice, or maize show that a *pair of allelic genes segregate from each other just as a pair of homologous chromosomes separate from each other during meiosis.*

THE STAGE OF MEIOSIS AT WHICH SEGREGATION OCCURS

It has been stated that the segregation of a pair of alleles in either a haploid or diploid organism derives from the fact that a pair of homologous chromosomes separate at anaphase I of meiosis. However, tetrad analysis, particularly in *Neurospora*, reveals that although the segregation of a pair of alleles is one to one as expected, they may not always segregate at anaphase I. An interesting feature of the life cycle of *Neurospora* (Fig. 2-6) is that the narrow character of the ascus results in an ordered alignment of the nuclei, and consequently the ascospores, in accordance with the position of the spindles during the meiotic and the postmeiotic divisions. The tetrads of *Neurospora* are accordingly *ordered* within the ascus. An analysis of segregation in this organism will reveal that a pair of alleles segregates from each other as predicted and will also reveal the order of their segregation with respect to all four chromatids of a given bivalent.

Wild-type *Neurospora* produces conidia that are pink in color. A mutant form, known as *albino* (*al*), has white conidia. When a cross is made with these two strains, tetrad analysis reveals that there is a one-to-one segregation of wild type and *albino*. However, when the ascospores are removed from the ascus *in order*, two classes of tetrads are observed. In one class, the order of genotypes within the ascus is +, +, *al*, *al*; in the other it is +, *al*, +, *al*. Segregation in both cases *is* one to one, and only the sequence of wild-type and mutant products within the ascus has changed. These two patterns of segregation can be understood by examining the events that take place during meiosis (Fig. 3-8). The sequence +, +, *al*, *al* results from the segregation of the alleles at the *first meiotic division* or, more precisely, when the homologous centromeres segregate at anaphase I. In a tetrad with the order +, *al*, +, *al*, the + and *al* alleles did not segregate at anaphase I, for at the first division apparently both a mutant and a wild-type allele went to each pole. However, this sort of segregation could occur at anaphase II. You will recall that chiasmata are formed during prophase of meiosis, and that they may represent crossovers or physical

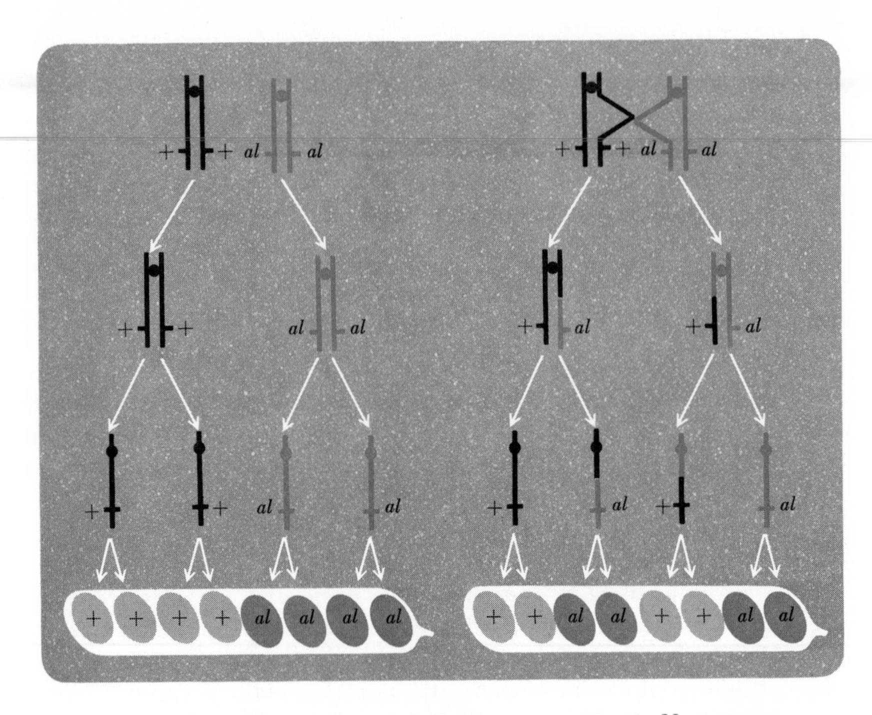

Fig. 3-8 First and second division segregation in Neurospora.

exchanges between nonsister chromatids. If a crossover were to occur *between the centromere and the locus* of *al* and its + allele, it can be seen from Fig. 3-8 that segregation of *al* from + would then occur at the second division. This segregation is caused by the fact that two nonsister chromatids have undergone an exchange during prophase, so that each pair of chromatids now carries a + and *al* allele. These alleles cannot segregate until the centromere divides at anaphase II and the sister chromatids separate.

Segregation can occur, therefore, at either of the meiotic divisions. Segregation at anaphase I or II will depend on whether or not a crossover has occurred between the locus of the allele in question and its centromere.

MULTIPLE ALLELES In considering the transmission of the genetic material, we have dealt with a pair of alleles, wild-type and mutant, at a specific locus in a chromosome. There are, however, many instances in which there appear to be a number of mutant alleles at the same locus, as for example, in the genetics of coat color in domestic rabbits. The wild-type rabbit is fully colored

Table 3-1 **Results of Crosses between Chinchilla, Himalayan, and Albino Rabbits**

P	chinchilla × Himalayan		chinchilla × albino		Himalayan × albino	
	↓		↓		↓	
F_1	chinchilla		chinchilla		Himalayan	
	↓		↓		↓	
F_2	chinchilla	Himalayan	chinchilla	albino	Himalayan	albino
	3 :	1	3 :	1	3 :	1

and has a brownish-gray coat. Distinct from the wild-type rabbit is a mutant known as *albino*, which lacks pigment. There are, in addition, a number of other mutant coat-color phenotypes of which two, *chinchilla* and *Himalayan*, will be considered here. Rabbits with the chinchilla phenotype are fully colored but, unlike wild-type rabbits, their hair is silver. Himalayan rabbits are white except for their extremities; the nose, tail, and forelegs are pigmented.

Homozygous strains of these rabbits are maintained by breeders, and the genetics of coat color has been investigated. When either homozygous chinchilla, Himalayan, or albino animals are crossed with homozygous wild type, the F_1 has the wild phenotype. When the F_1 from each of these crosses are inbred, the F_2 show a three-to-one segregation of wild type from mutant. These results indicate that chinchilla, Himalayan, and albino differ from wild type by a single gene mutation, and that each is recessive to the wild-type allele. There is nothing unusual about these results. When the mutants are crossed with each other, however, and their respective F_1 are inbred, the results show the form of inheritance characteristic of a pair of alleles (Table 3-1). In other words, there is a wild-type allele with three mutant alleles at the same locus. The wild-type allele is dominant over the three mutants; chinchilla is dominant over Himalayan and albino; and Himalayan is dominant over albino.

These alleles, all concerned with coat color, represent a series of *multiple alleles* at the same gene locus. Since each mutant allele is responsible for a different coat color, it may be concluded that each affects pigment synthesis in a different way, and that each allele is derived by a mutation that affects this function in a different way. We can exemplify this diagrammatically in the following manner. Assume that the coat-color gene occupies a site within the DNA which is delimited hypothetically by two vertical lines

A mutation may affect this site in a number of places, and the mutation wild type → chinchilla ($+ \rightarrow c^{ch}$) might be

------| ------c^{ch}------ |--------

Accordingly, Himalayan (c^h) and albino (c) may be

------|$\underline{c^h}$-----------|----- and ------|---------------$\underline{c}$|-----

respectively. These diagrams indicate that there may be more than one unit of mutation within a single gene locus, and that mutation at different sites within this locus leads to different alterations in the function of the same gene.

There are many genetic characteristics that are determined by a series of multiple alleles. Among them are the various blood groups of man and certain other vertebrates. For example, the ABO blood groups of man comprise a multiple-allelic series, and the presence of different alleles in this series can be distinguished immunologically. When blood obtained from two individuals is mixed, a clumping or *agglutination* of cells may or may not occur. The agglutination reaction occurs when serum from an individual of blood group A is mixed with the red blood cells of an individual of group B. The same reaction occurs in the reverse mixture of serum and blood cells (that is, serum from a group B individual and cells from a group A individual). A mixture of serum and cells from individuals of the same group does not give an agglutination reaction. A person of blood group A possesses an antibody (β) against group B blood cells. This antibody reacts with the antigen (B) in the red blood cells of group B individuals. Similarly, persons of group B have an antibody (α) against the antigen (A) present in individuals of the A blood group. Groups AB and O can also be distinguished immunologically. Group AB individuals have no antibodies in their serum, but possess the antigens A and B in the red blood cells. Group O blood is distinguished by the absence of antigens in the red blood cells and the presence of both the alpha and beta antibodies in the serum (Table 3-2).

Let us examine the inheritance of the four blood groups. It is relatively easy to obtain large samples of individuals for blood typing, and when classification is done on a familial basis, it is found that group A parents are of two sorts: those that have only group A progeny, and

Table 3-2 **The Agglutination Reactions Observed with the A, B, AB, and O Blood Groups**

Blood group	*Type of antibody*	*Types of red blood cells agglutinated*
A	β	B, AB
B	α	A, AB
AB	none	none
O	α and β	A, B, and AB

those that have group A and group O progeny. Similarly, group B parents may have either all group B progeny, or group B and group O progeny. The offspring of group O parents are group O, and the offspring of group AB parents may be A, B, or AB. Group A blood is due to a gene L^A, which controls the synthesis of the antigen *A*, and group

Table 3-3 **Parental Blood Groups and the Blood Groups of Their Offspring**

O × O	Possible parental genotypes	$ll \times ll$	—	—	—
	Phenotype of F_1	O	—	—	—
O × A		$ll \times L^AL^A$	$ll \times L^Al$	—	—
		A	A or O	—	—
O × B		$ll \times L^BL^B$	$ll \times L^Bl$	—	—
		B	B or O	—	—
O × AB		$ll \times L^AL^B$	—	—	—
		A or B	—	—	—
A × A		$L^AL^A \times L^AL^A$	$L^AL^A \times L^Al$	$L^Al \times L^Al$	—
		A	A	A or O	—
A × B		$L^AL^A \times L^BL^B$	$L^AL^A \times L^Bl$	$L^Al \times L^BL^B$	$L^Al \times L^Bl$
		AB	A or AB	B or AB	A, B, AB, or O
A × AB		$L^AL^A \times L^AL^B$	$L^Al \times L^AL^B$	—	—
		A or AB	A, B, or AB	—	—
B × B		$L^BL^B \times L^BL^B$	$L^BL^B \times L^Bl$	$L^Bl \times L^Bl$	—
		B	B	B or O	—
B × AB		$L^BL^B \times L^AL^B$	$L^Bl \times L^AL^B$	—	—
		B or AB	A, B, or AB	—	—
AB × AB		$L^AL^B \times L^AL^B$	—	—	—
		A, B, or AB	—	—	—

B blood is due to an allele L^B controlling the synthesis of the antigen *B*. Group O individuals have a third allele *(l)* in whose presence neither antigen is produced. Thus, group *A* individuals can be of two genotypes, homozygous L^AL^A and heterozygous $L^A\ l$. Group B individuals will be either L^BL^B or L^Bl, group AB individuals (who have both antigens) will be L^AL^B, and group O individuals will be *ll*. The various parental genotypes and possible offspring produced are shown in Table 3-3.

We can conclude from these observations that the alleles L^A and L^B do not interact, for AB individuals produce both antigens. Both alleles, however, appear to be dominant to the allele *l*, and group A and group B heterozygotes are immunologically indistinguishable from their respective homozygotes. Further analysis has revealed a variety of subgroups within the A and B groups, and several different alleles at this locus are known. The function of each allele at the ABO gene locus is different, and we assume that each of these functions arose by a mutation that affected the locus in a dissimilar way.

FURTHER READING

Iltis, H., 1932. *Life of Mendel*, translated by E. and C. Paul. New York: W. W. Norton & Company, Inc.

Mendel, G., 1966, "Experiments in plant hybridization," an English translation of "Versuche über Pflanzen Hybriden," reprinted in *Classic papers in genetics*, 1959. Peters, J. A. (ed.). Englewood Cliffs, N. J.: Prentice-Hall, Inc.

Srb, A. M., Owen, R. D., and Edgar, R. S., 1965. *General genetics*. San Francisco: W. H. Freeman and Company.

PROBLEMS

1. A mutant strain of *Chlamydomonas* having paralyzed flagella (*pf*) was crossed to the motile, wild-type strain and 100 tetrads were analyzed. Each tetrad gave a one-to-one ratio of paralysis to motility. Explain this result in terms of the behavior of a pair of homologous chromosomes during meiosis.

2. Instead of analysis of 100 tetrads, as in question 1, assume that one meiotic product was recovered at random from each of the 100 tetrads. What would be the ratio of paralyzed to motile?

3. A cross was made between two *Drosophila* that had the wild phenotype. Their progeny were found to have the same phenotype. A sample was taken of 200 of the progeny, each of which was crossed with a fly having a purple eye color. Half of the crosses gave only wild-type flies and the other half gave 50 percent wild-type and 50 percent purple-eyed progeny. What were the genotypes of the original pair of wild-type flies?

4. A cross was made between wild-type (+) *Neurospora* and a mutant strain that cannot synthesize the vitamin thiamin *(thi)*. Tetrad analysis showed a one-to-one segregation for +: *thi*. It was also found that 20 percent of the tetrads showed second-division segregation. How were these tetrads recognized and what are the events in meiosis that lead to second-division segregation?

5. Albinism in man is the absence of pigment from the hair, skin, and eyes. Determine from the pedigree given below whether it is a dominant or a recessive gene.

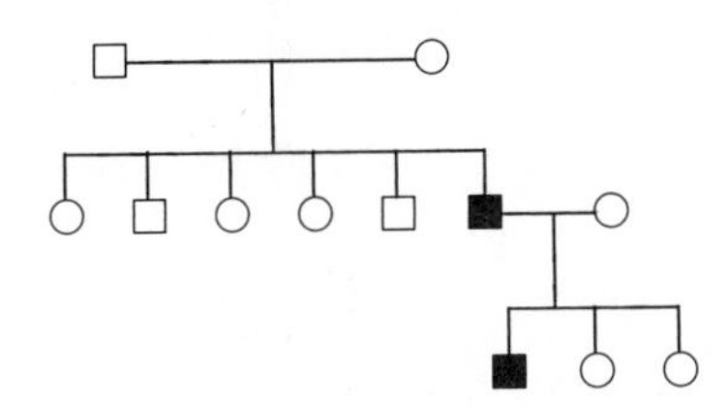

6. A rabbit cage contained one Himalayan female, one albino male, and one chinchilla male. The female had eight offspring: two Himalayan, four chinchilla, and two albino. Which rabbit was the paternal parent and what are the genotypes of the female and male parents and their offspring?

chapter 4

Independent Assortment

Since any given pair of alleles segregates in a one-to-one fashion, we must consider next what will occur when we deal with the segregation of two or more pairs of alleles in nonhomologous chromosomes. It was pointed out in Chapter 2 that during the first metaphase of meiosis any two pairs of nonhomologous chromosomes not only are independent of each other in the way in which they become aligned at the equator of the cell, but also they behave independently of each other throughout meiosis. This is observed when the members of each pair can be distinguished from one another as shown diagrammatically in Fig. 2-3G and H. If the homologous chromosomes of one pair are labeled *A* and *a* and those of the other pair *B* and *b*, we observe

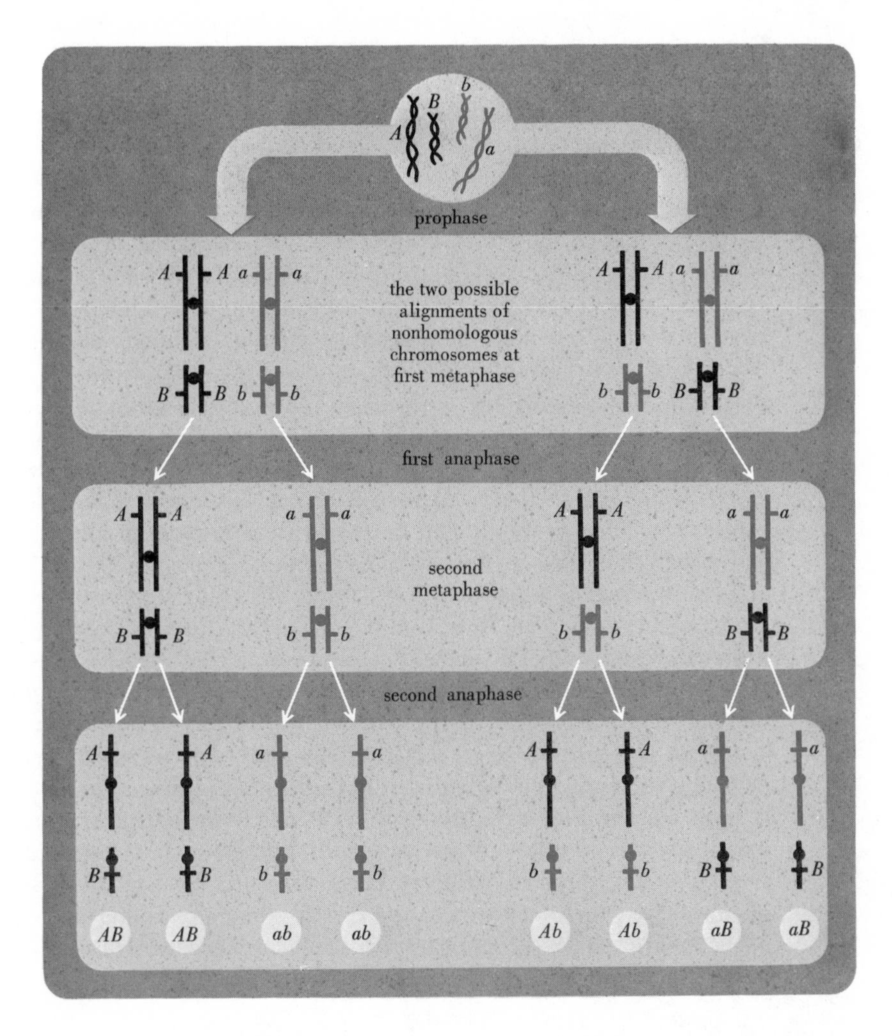

Fig. 4-1 *Independent assortment of two pairs of chromosomes at meiosis.*

that meiosis yields an equal number of the combinations *AB*, *Ab*, *aB*, and *ab* (Fig. 4-1).

INDEPENDENT ASSORTMENT–TETRAD ANALYSIS

If we now substitute two pairs of alleles, *A*,*a* and *B*,*b* for the hypothetical chromosomal markers used earlier, we can hypothesize that they too will behave independently, as shown in Fig. 4-1. The hypothesis can be tested by tetrad analysis.

For instance, a cross can be made between an *arginine*- requiring strain *(arg)* and an *acetate*-requiring strain *(ac)* of *Chlamydomonas*, and the tetrads can be analyzed by the technique described in the preceding chapter. The meiotic products are separated and grown to form colonies on medium containing both arginine and sodium acetate. The colonies are then replica plated to four different kinds of media: minimal medium, minimal medium plus arginine, minimal medium plus sodium acetate, and minimal medium plus arginine and sodium acetate.

The cross of the two mutant strains is symbolized as *arg* + × + *ac* and can be made using either strain as the plus or minus mating type. According to our hypothesis, four kinds of progeny are expected from this cross: *arg* +, + *ac*, + +, and *arg ac*. They can be distinguished from each other by the media on which they will or will not grow (Table 4-1).

Tetrad analysis of the cross gives the results shown in Table 4-2. It can be seen that three classes of tetrads are found. The first class consists of colonies derived from two sorts of meiotic products, *arg* + and + *ac*. Since two of the products have the genotype of one parent and two the genotype of the other parent, this class of tetrad is called a *parental ditype* or PD. Note that *each pair of alleles is segregating in a one-to-one fashion.* The second class of tetrad also shows a one-to-one segregation for each pair of alleles, but it consists of two new combinations of genotypes. Two of these new meiotic products are *arg ac* in genotype, and two are ++. This tetrad, which consists of two types of products each genotypically different from either of the original parents, is called a *nonparental ditype* or NPD. The two new genotypes are called *recombinant*, in contrast to the *parental* genotypes. Four different genotypes are found in the third class of tetrad. Two of these are the parental types (*arg* + and + *ac*) and two are the nonparental or recombinant types (*arg ac* and + +). A tetrad of this sort is called a *tetratype* or *T*. Again, each pair of alleles segregates one to one.

Table 4-1 **Types of Progeny Produced in the Cross of *arg* + × + *ac* in *Chlamydomonas* and the Media on Which They Will Grow**

(Growth is indicated by a plus sign, absence of growth by a minus sign)

Type of progeny	*Minimal*	*Minimal + arginine*	*Minimal + acetate*	*Minimal + arginine and acetate*
arg ac	−	−	−	+
+ *ac*	−	−	+	+
arg +	−	+	−	+
+ +	+	+	+	+

Table 4-2 **The Three Classes of Tetrads Produced in the Cross *arg* + × + *ac* in *Chlamydomonas***

TETRAD CLASS	*arg* + *arg* + + *ac* + *ac*	*arg ac* *arg ac* + + + +	*arg* + + + *arg ac* + *ac*	
NUMBER	71	69	95	Total = 235

Let us consider the origin of the three classes of tetrads. As in segregation, their origin derives from the behavior of the chromosomes during meiosis. Assuming that the two pairs of alleles are in *nonhomologous chromosomes*, then the PD and NPD tetrads arise as the result of the independent alignment of the chromosomes at metaphase I (Fig. 4-2A). Since the occurrence of the two different alignments is equally probable, then the ratio of PD:NPD tetrads will be one to one. The data given in Table 4-2 show that there are 71 PD and 69 NPD tetrads. This ratio does not differ markedly from a one-to-one ratio. A PD:NPD ratio of one to one is indicative of the *independent assortment* of two pairs of alleles. It means that the two pairs of alleles, just as the two nonhomologous pairs of chromosomes, behave independently of each other at meiosis.

Consider next the tetratype tetrads and their relation to the hypothesis of independent assortment. As stated in Chapter 3, segregation can occur at either the first or second division of meiosis. If it occurs at the second division for either of the two pairs of alleles, the tetrad formed will be a tetratype. This is shown in Fig. 4-2B, where second-division segregation has occurred for *arg* and its + allele and first division segregation for *ac* and its + allele. A tetratype tetrad will be formed regardless of which pair of alleles segregates at the second division. Work this out for yourself, as well as for the simultaneous occurrence of second-division segregation for both pairs of alleles.

The relation of tetratype tetrads to independent assortment can be seen when we examine the ratio of genotypes regardless of the tetrad class. The data given in Table 4-2 show that the four genotypes, *arg* +, + *ac*, + +, and *arg ac* are in a 1:1:1:1 ratio. Since tetratype tetrads, regardless of their number, consist of one of each of the four possible genotypes, they do not alter the 1:1:1:1 ratio. This ratio, as seen by examination of the number of genotypes, rather than the 1:1 ratio of PD:NPD tetrads, also shows that each of the alleles of a given pair can *assort independently with respect to any member of another allelic pair in a nonhomologous chromosome.*

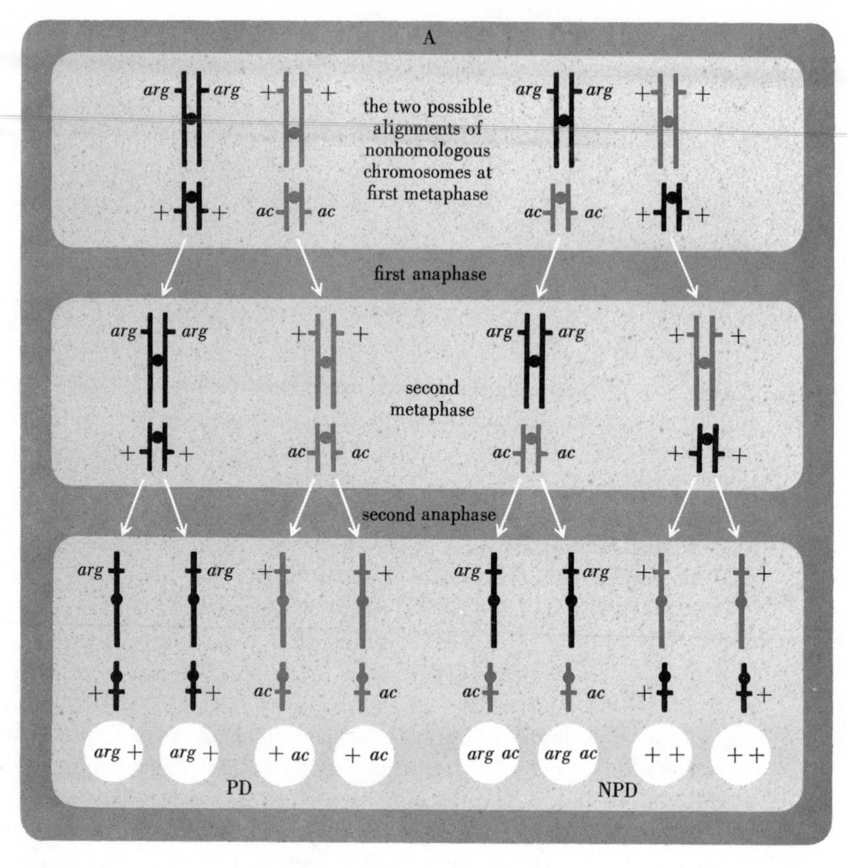

Fig. 4-2 *(A,* above, *and B,* facing.*) Independent assortment of two pairs of alleles in tetrads of* Chlamydomonas.

INDEPENDENT ASSORTMENT–STRAND ANALYSIS When we determined the genotypic ratio from the data in Table 4-2 without concern for tetrad class, we were essentially carrying out a strand analysis. Just as for segregation, the demonstration of independent assortment in a diploid such as *Drosophila* must be done by strand analysis.

A cross of a female *Drosophila*, homozygous for *vestigial*, to males that are homozygous for the recessive body-color mutation *ebony* (*e*) gives an F_1 that is wild in phenotype. We have already seen that *vg* is recessive, and the occurrence of a wild-type F_1 in the present cross indicates that *e* is also recessive. The F_1, therefore, is heterozygous for both *vg* and *e* and has the genotype $\frac{+}{vg}\frac{+}{e}$. These F_1 flies should produce equal numbers of four types of gametes, *vg* +, + *e*, + +, and *vg e*, if the two pairs of alleles assort independently. This statement can be proved by performing a test cross; that is, by crossing the F_1 flies to flies homozygous for *vg* and *e*. When this is done, four different kinds of flies are obtained in essentially equal numbers (Fig. 4-3).

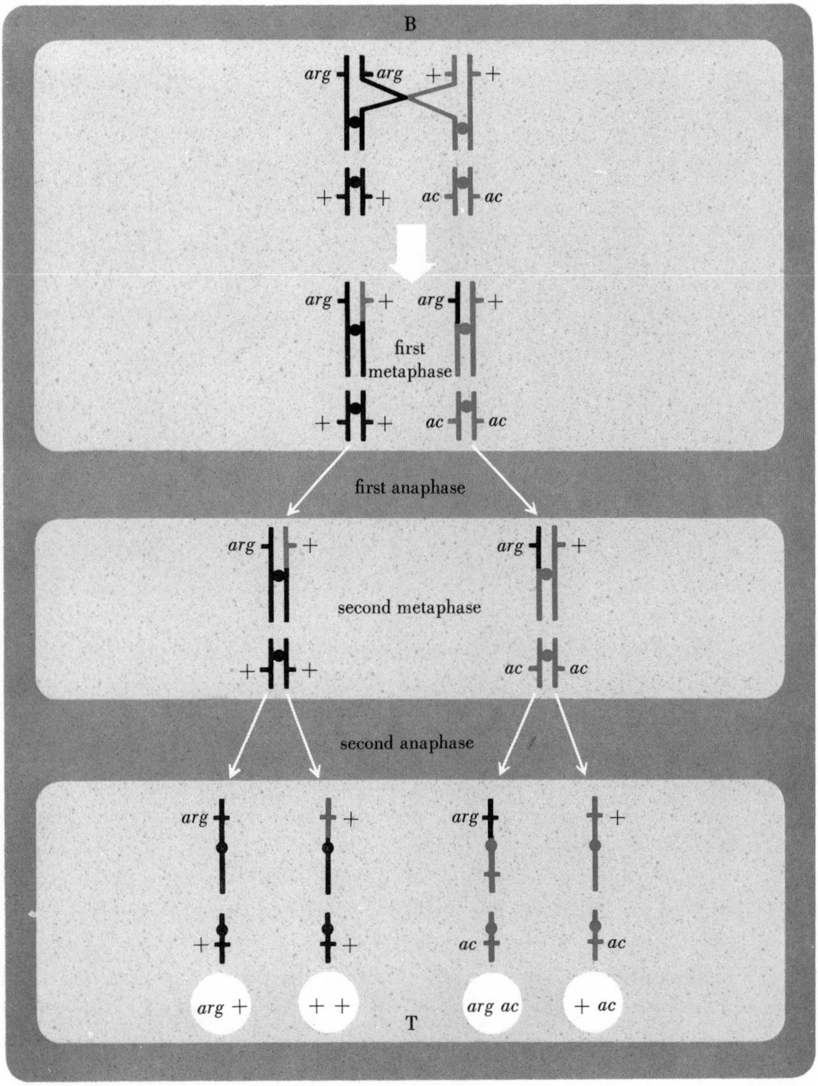

As in either tetrad or strand analysis with *Chlamydomonas*, there are four classes of phenotypes. Two classes are like the original parents: *vestigial* flies and *ebony* flies. Two classes exhibit nonparental or recombinant phenotypes: *wild type* and *vestigial-ebony*. The ratio of

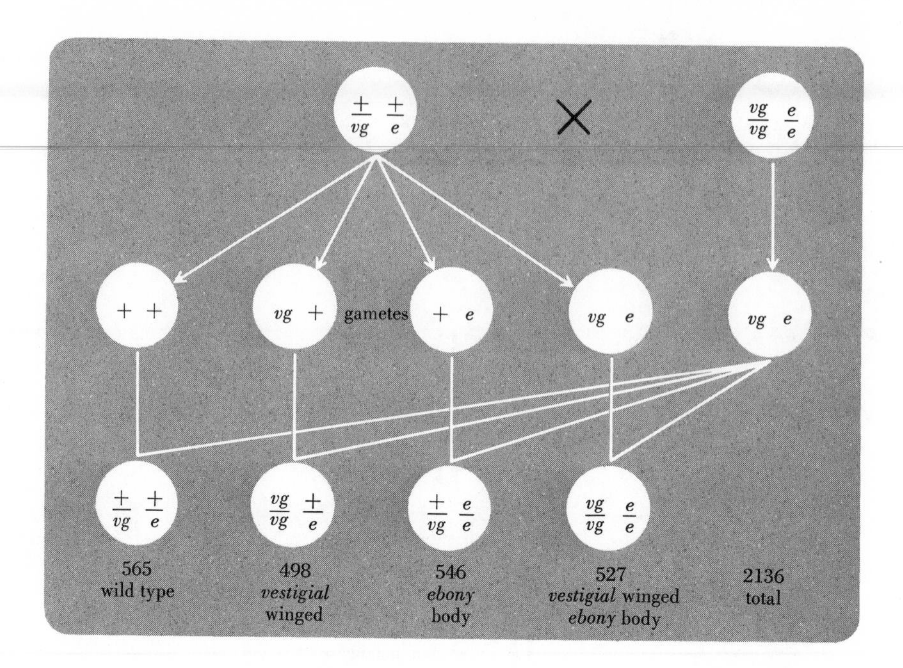

Fig. 4-3 *Independent assortment in a test cross of +/vg +/e* Drosophila.

the four classes is 1:1:1:1. Remember, however, that in the case of the diploid we are analyzing a random sample of meiotic products or strands. This is identical to classifying the genotypes from tetrads *without regard to tetrad class.*

A random sampling of gametes, or meiotic products, gives a ratio of flies that is 1 *vg* + : 1 + *e* : 1 + + : 1 *vg e*. There is, of course, no direct way of determining what proportion of gametes has arisen through second-division segregation of one or the other pair of alleles. We know from tetrad analysis, however, that the occurrence of tetratype tetrads does not alter the frequency of parental to nonparental genotypes.

With the test cross we have demonstrated that *vg* and *e* assort independently, as do *arg* and *ac* in *Chlamydomonas*. Thus, independent assortment in a diploid organism results from the random alignment of nonhomologous chromosomes at metaphase I of meiosis. In a diploid organism, heterozygous for two different pairs of alleles, an equal number of four types of gametes is produced at meiosis. In a haploid organism such as *Chlamydomonas* an equal number of four types of haploid vegetative cells is produced upon completion of meiosis. This is the Principle of Independent Assortment and we owe our knowledge of it to the work of Mendel.

There is an additional, though more indirect, way to demonstrate independent assortment in a diploid. For example, if the F_1 hetero-

Table 4-3 **The Phenotypes Obtained by Inbreeding *Drosophila* of the Genotype $\frac{+}{vg}\frac{+}{e}$**

Phenotype	*Number*
+	2834
vg	920
e	951
vg e	287
Total	4992

zygotes $\frac{+}{vg}\frac{+}{e}$, are inbred rather than test crossed, four phenotypes are obtained as before, but this time in different proportions. A typical experiment in which the F_1 are inbred gives the F_2 data shown in Table 4-3. The ratio of phenotypes is approximately 9:3:3:1 or 9/16 : 3/16 : 3/16 : 1/16. The wild-type flies are about nine times more frequent than the *vestigial-ebony* flies, and the *vestigial* flies and the *ebony* flies are about three times more frequent than *vestigial-ebony* flies.

The four phenotypes, *vestigial*, *ebony*, *wild type*, and *vestigial-ebony* are distributed among nine different genotypes, as shown in Table 4-4. The nine different genotypes are derived from the random combinations of the four sorts of gametes produced in equal numbers by each of the heterozygous parents. Proof of the existence of the nine

Table 4-4 **The Genotypes Obtained by Inbreeding *Drosophila of the Genotype* $\frac{+}{vg}\frac{+}{e}$**

Eggs	*Sperm*			
	+ +	+ *e*	*vg* +	*vg e*
+ +	$\frac{+}{+}\frac{+}{+}$	$\frac{+}{+}\frac{e}{+}$	$\frac{vg}{+}\frac{+}{+}$	$\frac{vg}{+}\frac{e}{+}$
+ *e*	$\frac{+}{+}\frac{+}{e}$	$\frac{+}{+}\frac{e}{e}$	$\frac{vg}{+}\frac{+}{e}$	$\frac{vg}{+}\frac{e}{e}$
vg +	$\frac{+}{vg}\frac{+}{+}$	$\frac{+}{vg}\frac{e}{+}$	$\frac{vg}{vg}\frac{+}{+}$	$\frac{vg}{vg}\frac{e}{+}$
vg e	$\frac{+}{vg}\frac{+}{e}$	$\frac{+}{vg}\frac{e}{e}$	$\frac{vg}{vg}\frac{+}{e}$	$\frac{vg}{vg}\frac{e}{e}$

genotypes can be obtained by test crossing the F_2 flies singly with $\frac{vg}{vg}\ \frac{e}{e}$ flies. We can predict from the genotypes shown in Table 4-4 that among the wild-type flies, 1/9 are homozygous for both wild-type alleles, 2/9 are heterozygous for *vg*, 2/9 are heterozygous for *e*, and 4/9 are heterozygous for both *vg* and *e*. Similar predictions can be made for the genotypes of the *vg* and *e* flies of the F_2. Work out these test crosses to prove for yourself that the predicted genotypes are found in the proportions expected.

THE CHI–SQUARE TEST The results of crosses involving one or more pairs of alleles can be predicted, and the prediction tested by the appropriate crosses. Since, in the case of diploids, strand analysis depends upon a random sampling of the meiotic products, it is necessary to have some means for determining whether or not the data derived from either a test cross or from inbreeding the F_1 heterozygotes are in agreement with a given prediction or hypothesis. Thus, we can ask whether the F_2 data (Table 4-3) obtained by inbreeding the $\frac{+}{vg}\ \frac{+}{e}$ flies fit a ratio of 9:3:3:1. A statistical test known as chi-square (χ^2) can be used to test the "goodness of fit" of these data to any given ratio.

The χ^2 test involves, first, determining a predicted ratio and, second, establishing how closely the observed data fit this ratio. For a detailed discussion of the χ^2 test see Frederick C. Mills, *Statistical Methods*, 3rd ed. (Holt, Rinehart and Winston, 1955). The χ^2 test is made by ascertaining the probability that the deviation of the observed ratio from the predicted ratio is due to chance, and not to some other factor such as experimental conditions, sampling, or even the wrong hypothesis. The usual statistical procedure is to establish an arbitrary criterion for a deviation significant from what would be expected from chance alone. If the *probability (P)* of obtaining the observed ratio by chance alone is equal to or less than five in a hundred ($P = 0.05$) then the deviation between the expected and the observed ratio is considered significant. If the probability is one in a hundred or less ($P = 0.01$) then the deviation is highly significant. When the P value is greater than 0.05, the deviation is not considered statistically significant and can be expected on the basis of chance alone.

A χ^2 test for the data given in Table 4-3 is worked out as follows:

1. *The hypothesis:* That the observed phenotypic ratio is in accord with a predicted phenotypic ratio of 9/16 + : 3/16 *vg* : 3/16 *e* : 1/16 *vg e*.

2. *Expected or predicted ratio:* Dividing the total number of 4992 by 16, the expected number of *vg e* flies is determined to be 312. The expected number of wild-type flies is therefore 9 × 312 or 2808, and the expected number of *vg* flies and of *e* flies is accordingly 3 × 312 or 936.

3. *Deviations between observed and expected ratios and the calculation of* χ^2:

	+	*vg*	*e*	*vg e*	Total
Observed (*O*)	2834	920	951	287	4992
Expected (*E*)	2808	936	936	312	4992
$D\ (O-E)$	26	16	15	25	–
D^2	676	256	225	625	–
D^2/E	0.24	0.27	0.24	2.00	–

$\Sigma\ D^2/E = \chi^2 = 2.75 \quad df = 3 \quad P = 0.50–0.30$

4. *The P value:* Tables of χ^2 values have been prepared from which it is possible to determine the corresponding P values. A part of such a table is given in Table 4-5.

The relationship between χ^2 and P is a function of the number of *degrees of freedom.* If n is the total number of classes of different genotypes, the degrees of freedom *(df)* usually equal $n - 1$. The larger the number of classes, the greater are the degrees of freedom. If there are four classes, the degrees of freedom will be equal to 3. In other words, with four classes of phenotypes, when the expected numbers for three are calculated from a given total number of individuals, the fourth number is set. In the foregoing example there are four classes, making a total of 4992. The expected values for the first three add up to 4680. This means that the fourth class must equal 312.

When $\chi^2 = 2.75$ and there are 3 degrees of freedom, the P value obtained from Table 4-5 is greater than 0.30 and less than 0.50. Thus, 30 to 50 times out of 100 we could expect chance deviations equal to or less than those observed. Since the value of P is greater than 0.05, the observed results for a 9:3:3:1 ratio are in good agreement with those to be expected for the independent assortment of two pairs of alleles.

GENERAL CASE FOR INDEPENDENT ASSORTMENT IN A DIPLOID

Let us establish a general case for independent assortment in a diploid for a cross in which doubly heterozygous individuals, $\frac{+}{a}\frac{+}{b}$, are inbred. Although there is no short

Table 4-5 Table of χ^2 **(Chi-square)**

n	*P* = 0.99	0.98	0.95	0.90	0.80	0.70
1	0.000157	0.00628	0.00393	0.0158	0.0642	0.148
2	0.0201	0.0404	0.103	0.211	0.446	0.713
3	0.115	0.185	0.352	0.584	1.005	1.424
4	0.297	0.429	0.711	1.064	1.649	2.195
5	0.554	0.752	1.145	1.610	2.343	3.000
6	0.872	1.134	1.635	2.204	3.070	3.828
7	1.239	1.564	2.167	2.833	3.822	4.671
8	1.646	2.032	2.733	3.490	4.594	5.527
9	2.088	2.532	3.325	4.168	5.380	6.393
10	2.558	3.059	3.940	4.865	6.179	7.267

Abridged from Table II of Fisher & Yates: *Statistical Tables for Biological, Agricultural and Medical Research* published by Oliver & Boyd Ltd., Edinburgh, and by permission of the authors and publishers.

method for deriving the genotypes and their ratios among the progeny produced, it can easily be seen from the information in Table 4-4 that a homozygous genotype will represent 1/16 of the total progeny, whereas a genotype heterozygous for one pair of alleles will be 2/16 of the total progeny. The genotype heterozygous for both pairs of alleles will be 4/16 of the total progeny.

There is, however, a simple method for determining phenotypic ratios. In the example of inbreeding $\frac{+}{vg}\frac{+}{e}$ flies, *vg* segregates from its + allele in a 3:1 ratio, and the same is true for *e* and its + allele. If the two pairs are assorting independently we can determine the probability of their combinations as follows:

3 out of 4 F_2 flies have wild-type wings and
1 out of 4 F_2 flies have *vestigial* wings;
3 out of 4 F_2 flies have wild-type body color and
1 out of 4 F_2 flies have *ebony* body color.

Therefore, the probability of occurrence of all possible combinations is as follows:

Wild-type wing and wild-type body color is $3/4 \times 3/4 = 9/16$
Wild-type wing and *ebony* body color is $3/4 \times 1/4 = 3/16$
Vestigial wing and wild-type body color is $1/4 \times 3/4 = 3/16$
Vestigial wing and *ebony* body color is $1/4 \times 1/4 = 1/16$

This method can be used to determine the phenotypic ratio when inbreeding individuals that are heterozygous for three pairs of alleles,

Table 4-5 continued

0.50	*0.30*	*0.20*	*0.10*	*0.05*	*0.02*	*0.01*
0.455	1.074	1.642	2.706	3.841	5.412	6.635
1.386	2.408	3.219	4.605	5.991	7.824	9.210
2.366	3.665	4.642	6.251	7.816	9.837	11.345
3.357	4.878	5.989	7.779	9.488	11.668	13.277
4.351	6.064	7.289	9.236	11.070	13.388	15.086
5.348	7.231	8.558	10.645	12.592	15.033	16.812
6.346	8.383	9.803	12.017	14.067	16.622	18.475
7.344	9.524	11.030	13.362	15.507	18.168	20.090
8.343	10.656	12.242	14.684	16.919	19.679	21.666
9.342	11.781	13.442	15.987	18.307	21.161	23.209

or in fact when any number of allelic pairs is involved and when dominance is complete. When there is incomplete dominance for each allelic pair the genotypic and phenotypic ratios are identical, just as we saw in the previous chapter when we dealt with the inheritance of flower color in the snapdragon. Several problems involving this sort of inheritance will be found at the end of this chapter.

In this chapter, as well as in the previous chapter, we have seen that the units of genetic material called genes are transmitted as if they were particulate entities, and that their transmission from one generation to the next is exactly parallel to the behavior of the chromosomes at meiosis. It has been seen, for example, that a heterozygote ($+/a$) produces two kinds of meiotic products, a and $+$, in equal numbers, by virtue of the segregation of homologous chromosomes at meiosis. When two or more pairs of alleles are involved, $\frac{+}{a}\frac{+}{b}$, it has been seen that four kinds of meiotic products ($a\,+$, $+\,b$, $+\,+$, and $a\,b$) are formed in equal numbers as a result of the independent assortment of nonhomologous chromosomes during meiosis.

PROBLEMS

1. In *Chlamydomonas* a mutant gene *arg* assorts independently of a mutant gene *pf*. What are the types of tetrads expected from this cross? How many different genotypes will there be and in what ratio?

2. The tetrads formed in *Neurospora* are ordered as a result of meiosis in the ascus (see p. 43 and Fig. 2-6). Assume that one mating type of this organism carries the mutant gene a and the wild-type allele of mutant gene b, while the other mating type carries the wild-type allele of the mutant gene a and the mutant gene b. A cross, $a\,+ \times +\,b$, gave the following kinds and numbers of ordered tetrads:

A	B	C	D	E	F
a +	a b	a +	a b	a +	+ +
a +	a b	a b	a +	+ +	a +
+ b	+ +	+ +	+ b	a b	+ b
+ b	+ +	+ b	+ +	+ b	a b
134	132	105	108	11	10

(a) Show how the data reveal that a and b are assorting independently.
(b) Classify each tetrad by name and show with diagrams the *simplest* origin for each. Indicate for each tetrad whether segregation occurs at either the first or second division for the a locus and b locus.
(c) Instead of analyzing 500 tetrads, assume that one ascospore was recovered at random from each of 1000 asci. How many genotypes would you find and how many of each kind?

3. A cross between two wild-type flies gave progeny all of which were wild type. When they were test crossed to $\frac{vg\ e}{vg\ e}$ flies the following results were obtained:

(a) 1/4 of the test crosses gave wild type, *vestigial ebony*, *vestigial*, and *ebony* in a 1:1:1:1 ratio.
(b) 1/4 of the test crosses gave all wild type.
(c) 1/4 of the test crosses gave *vestigial* and wild type in a 1:1 ratio.
(d) 1/4 of the test crosses gave *ebony* and wild type in a 1:1 ratio.

What are the genotypes of the original pair of wild-type flies?

4. Assume that the gene r and its + allele show incomplete dominance with respect to flower color such that +/+ is red, +/r is pink, and r/r is white and that the gene s and its allele show incomplete dominance with regard to seed color such that +/+ has red-black seeds, +/s pink seeds, and s/s white seeds. Give the phenotypes and their ratios expected among the progeny in the following crosses:

(a) $\frac{+\ s}{+\ s} \times \frac{r\ +}{r\ +}$

(b) $\frac{+\ +}{r\ s} \times \frac{+\ +}{r\ s}$

(c) $\frac{+\ +}{r\ s} \times \frac{r\ s}{r\ s}$

(d) $\frac{r\ +}{r\ s} \times \frac{+\ s}{r\ s}$

5. Two different true-breeding strains of sweet peas having white flowers were crossed. The flower color of the F_1 was purple. When the F_1 was inbred 9/16 of the progeny had purple flowers and 7/16 had white flowers. Explain these results in terms of what you have learned regarding independent assortment.

chapter **5**

Sex-linked Inheritance

Segregation and independent assortment, which result from the behavior of the chromosomes at meiosis, are essential features of the chromosome theory of inheritance. The theory gains additional support from other genetic phenomena. Among them is a form of inheritance in which the transmission of the genetic material is correlated with the sex of the parents. This form of inheritance, called *sex-linked inheritance*, is known in many different diploid organisms, and it shows that the inheritance of specific genes is correlated with the inheritance of a specific chromosome.

If we examine the chromosomes of *Drosophila melanogaster* we note that the males and females differ in the morphology

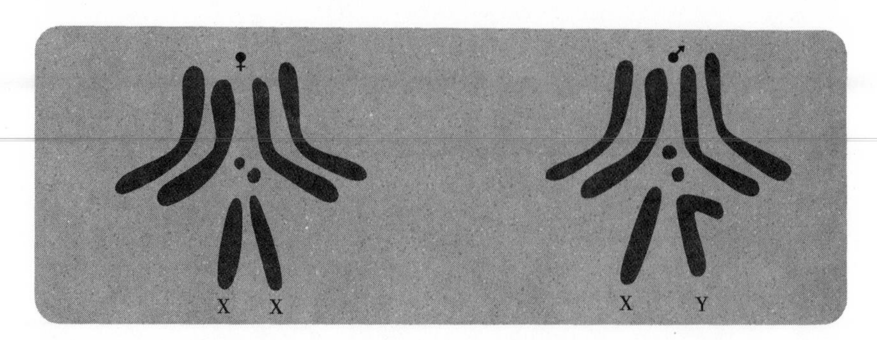

Fig. 5-1 X and Y chromosomes of Drosophila. *(Redrawn from Altenburg,* Genetics. *New York: Holt, Rinehart and Winston, 1957.)*

of one chromosome pair (Fig. 5-1). Both sexes possess a pair of small, dotlike chromosomes and two pairs of larger chromosomes in which the centromere is located in a median position. These three pairs of chromosomes are called *autosomes.* In females the fourth pair, the *X chromosomes* or sex chromosomes, have a nearly terminal centromere. The males have only one X chromosome. Paired with the X chromosome in the male is a morphologically different chromosome called the *Y chromosome.* Thus, females have three pairs of autosomes (3A) and two X chromosomes (XX), and males have three pairs of autosomes (3A) and an X and a Y chromosome (XY).

Sex in *Drosophila* appears to depend upon the presence of a pair of X chromosomes in females, or a single X chromosome and a Y chromosome in males. However, an occasional fertilization can lead to the production of flies having one X chromosome but no Y chromosome; such XO flies are males. Flies have also been obtained with two X chromosomes and a Y chromosome; these flies are females. The Y chromosome in *Drosophila,* therefore, plays no role in the determination of sex. In fact, it is known that femaleness depends upon the presence of two X or sex chromosomes, and two of each kind of the three autosomes. Maleness, on the other hand, is due to the presence of three pairs of autosomes and only one X chromosome. Sex determination in man follows a different pattern. It is known that the presence of a Y chromosome determines maleness and its absence femaleness. Exceptional individuals have been found who are XO, and they are females, whereas XXY exceptions are males.

The inheritance of the X chromosome in *Drosophila* follows a specific pattern (Fig. 5-2). A male *Drosophila* transmits his X chromosome only to his female offspring. His X chromosome is not transmitted to males until the *next generation.* A female *Drosophila,* on the other hand, transmits an X chromosome to both her male and female offspring.

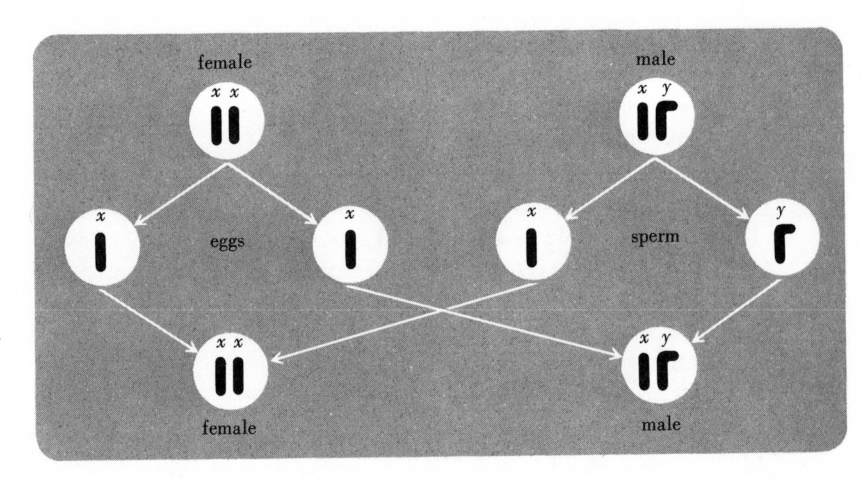

Fig. 5-2 Inheritance of X and Y chromosomes in Drosophila. *(After Altenburg.)*

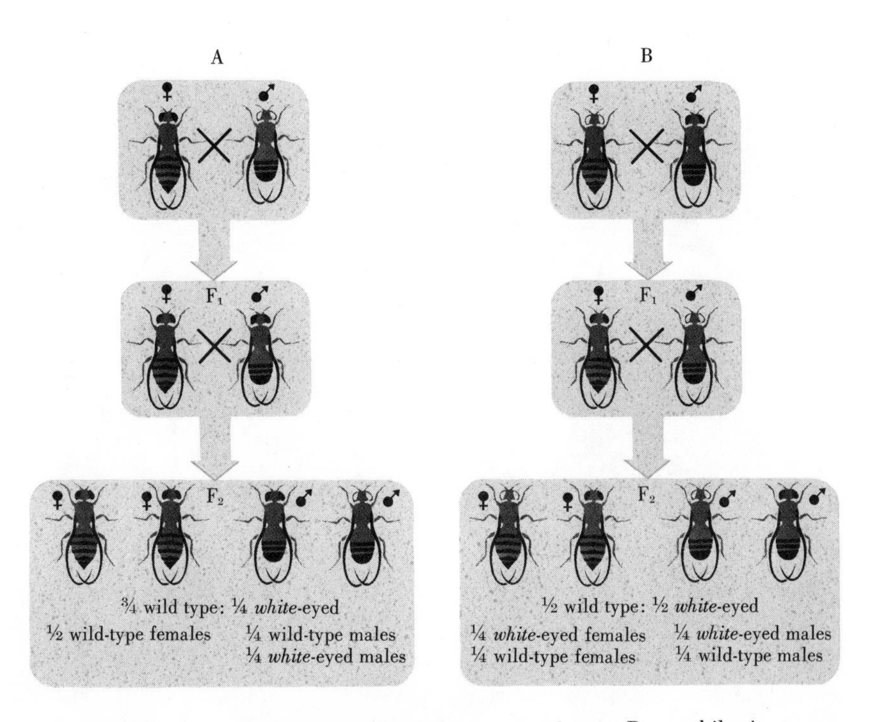

Fig. 5-3 Sex-linked inheritance of white eye color in Drosophila *in terms of phenotypes.*

You will note, therefore, that a male always inherits his X chromosome from his female parent, and his X chromosome is transmitted to his grandsons only through his female offspring. Any genetic traits that are transmitted in this fashion are *sex-linked.*

Among the numerous eye color mutants of *Drosophila* is one called *white (w).* Flies with the *white* phenotype have no eye pigments. If a cross is made between wild-type females and *w* males, the F_1 is wild type, and therefore, the *white* phenotype is recessive. When the F_1 is inbred, three quarters of the flies possess the wild phenotype and one fourth possess the *white* phenotype. This result appears to be a straightforward example of the segregation of a pair of alleles, *white* from its + allele. However, when the F_2 flies are classified for both eye color and sex it is found that all of the females possess the wild phenotype, but only half of the males are wild type, while the other half have *white* eyes (Fig. 5-3A). When the reciprocal cross is made—that is, *white* females crossed to wild-type males—the F_1 is composed of two different phenotypes: wild-type females and *white* males. When this F_1 is inbred, half of the progeny are wild in phenotype, and half of

Fig. 5-4 *(A,* below, *and B,* facing.*) Inheritance of white eye color in terms of genotypes and the X chromosome.*

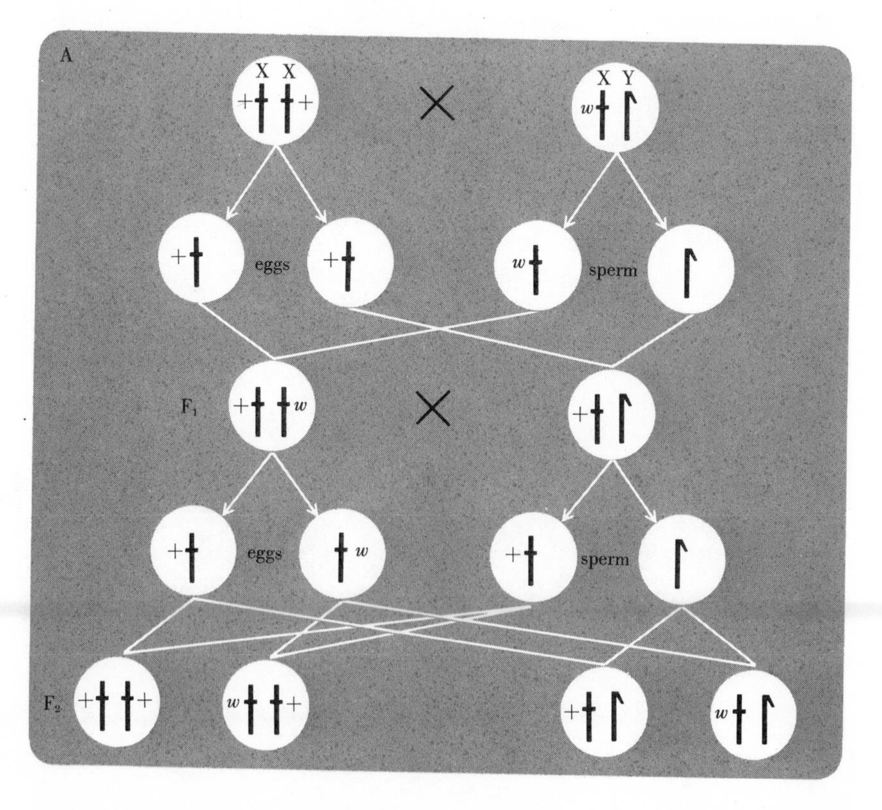

them are *white*. When classified as to sex and eye color, half of the females and half of the males are *white*. The remaining flies, both males and females, are wild type (Fig. 5-3B). These reciprocal crosses show a correlation between the sex of the individuals and the appearance of the *white* phenotype. In all of the examples considered heretofore, reciprocal crosses gave identical results with respect to sex in both the F_1 and F_2 generations.

In the cross shown in Fig. 5-3A (wild-type females × *white* males) all of the F_1 progeny have the wild phenotype, as might be expected. However, in cross B of the same figure (*white* females × wild-type males), *only* F_1 females have the wild phenotype whereas *all* of the F_1 males have the *white* phenotype. Referring again to Fig. 5-1, we see that the inheritance of *w* is in some way *linked* to *sex*, since if the female parent is *white* all of her male progeny are *white*. If the female parent is wild type, all of her offspring are wild type, and *white* is not seen until these progeny are inbred, and then the *white* phenotype appears only in the males.

In *Drosophila* the transmission of *w* and its wild-type allele (Fig. 5-3) can now be explained by assuming that *w* is located in the X chromosome. Figure 5-4A and B illustrates the pattern of transmission of the

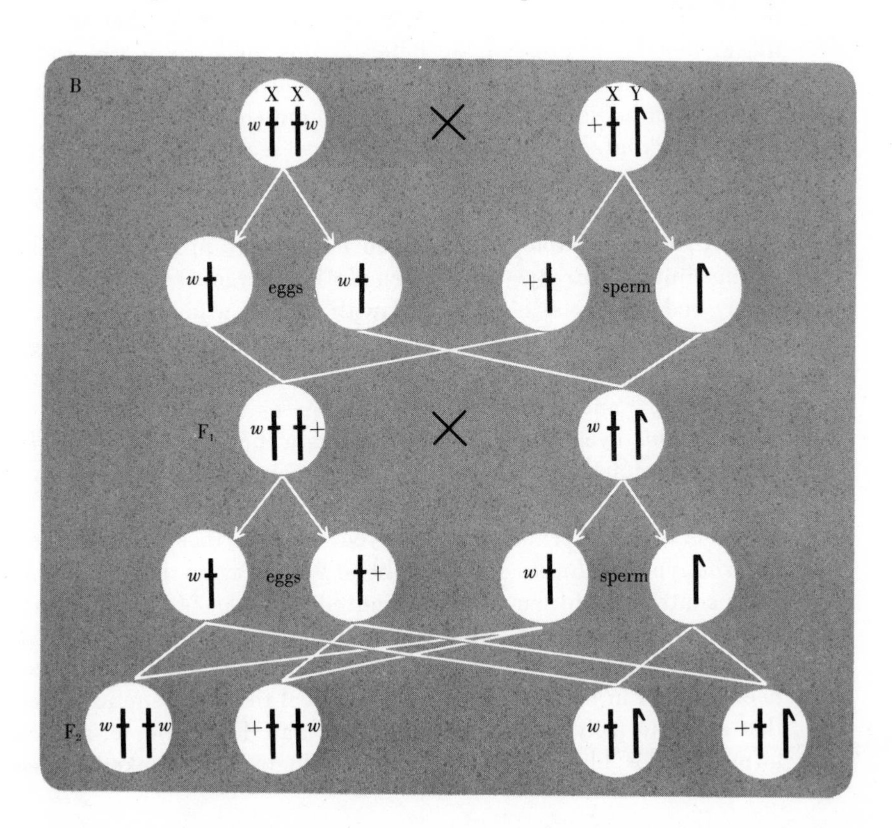

two alleles and of the X chromosome. The inheritance of the *w* gene is clearly sex-linked.

The F_1 males in both crosses inherited their X chromosomes from their female parents, and the male parents transmitted their X chromosomes only to their female offspring. In cross A, the F_1 males are wild type since their female parents produced gametes bearing an X chromosome with the + allele of *w*. In cross B, the F_1 males are *white* since the female parents produced gametes in which the X chromosome carried the *w* allele. Since *w* is recessive, the female progeny in both crosses are wild type. They received one X chromosome from a wild-type parent and one from a parent with the *white* phenotype. Because the males of cross B are *white*, the Y chromosome does not appear to carry an allele of *w*. In fact, it has been concluded that the Y chromosome is genetically inert with respect to most of the gene loci in the X chromosome of *Drosophila*.

With respect to sex-linked genes, male *Drosophila* are neither homozygous nor heterozygous since there are no allelic counterparts in the Y chromosome. Therefore, the males are referred to as being *hemizygous* for sex-linked genes.

Figure 5-4A and B shows the predicted genotypes of F_1 and F_2 flies of the reciprocal crosses. It is seen, for example, that one half of the F_2 females in cross A are heterozygous for *w*. Similarly, the wild-type F_2 females in cross B are heterozygous. The genotypes of these females can be verified by crossing them individually to *w* males. Work this out for yourself with diagrams. Is it necessary to cross these females to *w* males in order to test the predictions, or will males of any genotype serve the same purpose?

Another example of sex-linked inheritance in *Drosophila* is that of the mutant *Bar* eye *(B)*. A female homozygous for *B* has an eye highly reduced in size compared with that of wild-type flies. A *Bar*-eyed male has a phenotype similar to that of a *B/B* female. A heterozygous *Bar*-eyed female (*B*/+) has a phenotype intermediate between that of a +/+ and a *B/B* female. Two crosses will illustrate the inheritance of *Bar* eyes (Fig. 5-5). We see again the typical pattern of sex-linked inheritance in which the male genotype, with respect to *Bar*, is dependent on the genotype of his female parent. Thus, the male offspring of a *B/B* female are *Bar*-eyed, whereas those of a *B*/+ female are wild type and *Bar* in equal numbers. We see again that the X chromosome contributed by a male is transmitted only to his female offspring and that it does not appear in male progeny until the following generation.

The inheritance of a variety of sex-linked characteristics has been studied in man, and the studies reveal that the genetic principles that apply to fruit flies apply equally to man. For example, a frequent form of color blindness is well known to be a sex-linked recessive

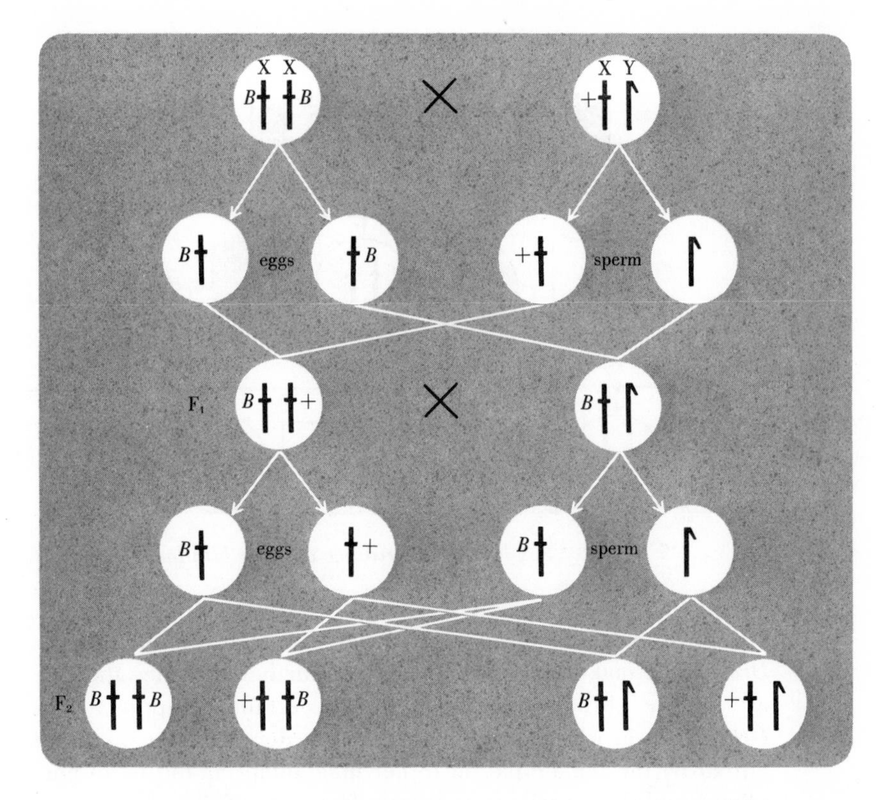

Fig. 5-5 The sex-linked inheritance of Bar *eye in* Drosophila.

characteristic in man. A pedigree of color blindness is given in Fig. 5-6. In the first generation a normal female, who has a color-blind brother, married a color-blind male. Of their three children, one male and one female were color blind, and one female was normal. The male offspring married a female with normal color vision and all their progeny had normal color vision. His color-blind sister married a man with normal color vision, and all of their male children were color blind, while their female child had normal color vision. When color blindness in this pedigree is analyzed in terms of the inheritance of the X chromosome, the sex-linked nature of the trait becomes evident (Fig. 5-6).

One of the significant features of sex-linked inheritance lies in the direct correlation between the transmission of a given chromosome, the X chromosome, and the transmission of specific hereditary traits. This correlation receives further support when one examines sex-linked inheritance among organisms in which the male, rather than the female, is XX. Such a situation exists in the birds and butterflies. An example of sex-linked inheritance in birds is given in Fig. 5-7.

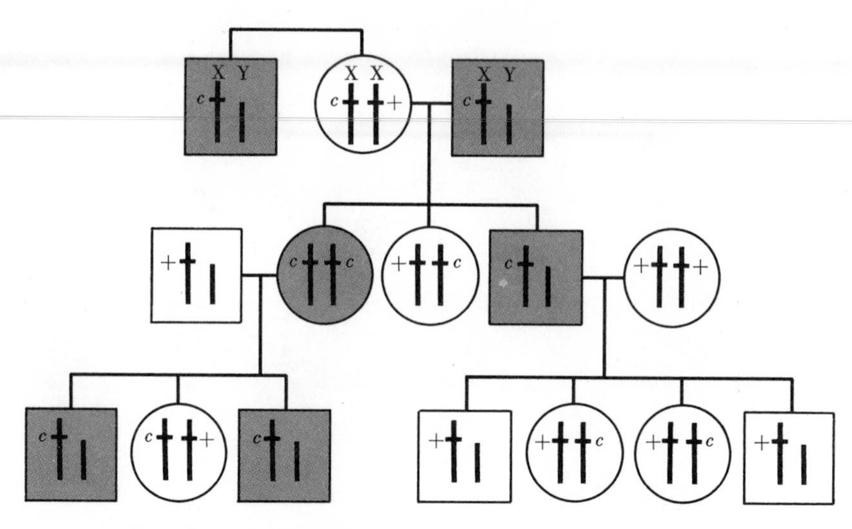

Fig. 5-6 Sex-linked inheritance of color blindness in man.

The existence of an exceptional case of sex-linked inheritance in *Drosophila* makes it possible to obtain further evidence that the genes are in the chromosomes. There are certain exceptional strains of *Drosophila* in which genes, sex-linked in normal strains, *are not transmitted* by the female parent to her male offspring but to all her *female* offspring. When a wild-type female from one of the exceptional strains

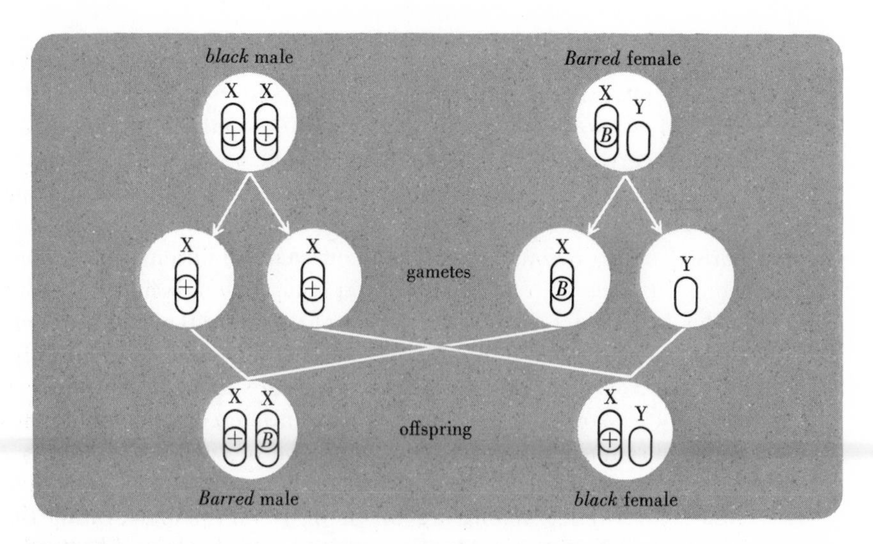

*Fig. 5-7 Sex-linked inheritance in chickens (*Barred *vs.* black *feathers). (After Altenburg.)*

is crossed to a *Bar*-eye male, the F_1 consists of wild-type females and *Bar*-eye males. This result, which is opposite to the normal case shown in Fig. 5-5, must be interpreted on the basis of males inheriting their X chromosome from their male parent rather than from their female parent.

Cytological examination of the chromosomes of the females from the exceptional strain reveals that their two X chromosomes, instead of being separate as is normally the case, are joined at their centromeres. These females with *attached X chromosomes* ($\widehat{XX}$) also have a Y chromosome. Two kinds of eggs are produced by these females: eggs with attached X chromosomes and eggs with a Y chromosome. An egg bearing an attached X chromosome, fertilized by a sperm bearing a Y chromosome will develop into a normal female, since she has two X chromosomes and the Y chromosome plays no role in sex determination. If this egg is fertilized by an X-chromosome-bearing sperm, however, the result is a *triplo-X* female. Such a female is usually inviable. If it does survive, it is readily recognized by a number of distinguishing characteristics. Fertilization of an egg bearing a Y chromosome by a sperm bearing an X chromosome results in a normal male. When this egg is fertilized by a sperm bearing a Y chromosome, the resulting zygote is inviable.

A cross between an attached-X female and a *Bar*-eye male is shown in Fig. 5-8. Four classes of offspring are possible: triplo-X

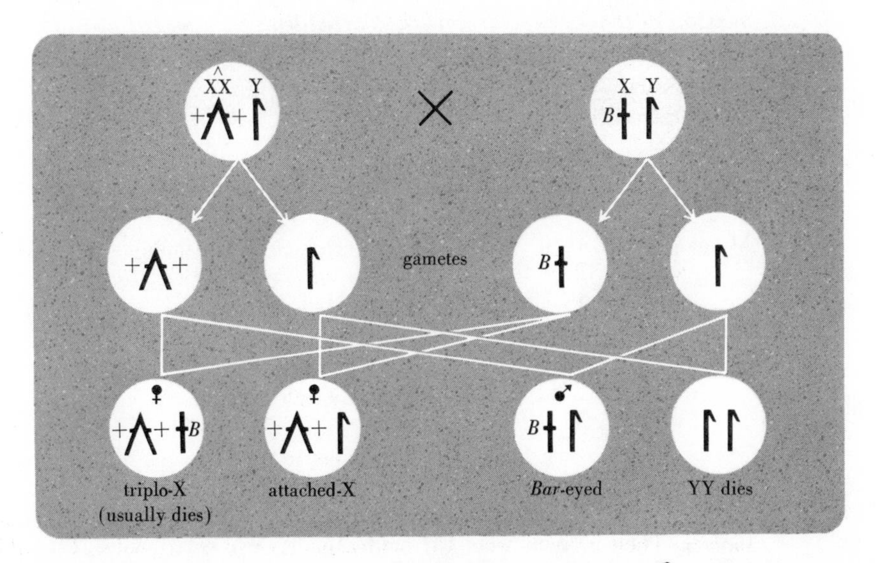

Fig. 5-8 *Results of a cross in* Drosophila *between an* $\widehat{XX}$ *female and a* Bar-*eye male.*

females with wild-type eyes, *Bar*-eye males, $\widehat{XX}$ females with wild-type eyes, and animals with two Y chromosomes but no X chromosome. These last are inviable, and the triplo-X females, as previously mentioned, usually die. Therefore, only two types of progeny are obtained: wild-type $\widehat{XX}$ females and Bar-eye males.

In this experiment there is a direct relation between a visible cytological chromosomal abnormality and the transmission of a specific hereditary trait. The *B* gene is present in the male parents and not in the females, and since the male offspring are *Bar* eyed they must have received the gene, and thus their X chromosome, from their male parent. Furthermore, since the X chromosomes of the female remain attached through meiosis, eggs with two rather than a single X chromosome are formed. Their fertilization can lead only to the development of females.

Sex-linked traits are inherited according to the manner in which the sex chromosomes are inherited, thus providing evidence to support the assumption that genes are indeed located within the chromosomes. Furthermore, the inheritance of a pair of sex-linked alleles follows the same principle of segregation shown for all alleles.

We have examined three sorts of evidence that support the chromosome theory of inheritance. First, *the separation of a pair of homologous chromosomes at meiosis is correlated with the segregation of a pair of alleles.* Second, *the random alignment of chromosomes at metaphase I of meiosis has its genetic counterpart in the phenomenon of independent assortment.* Third, *sex-linked inheritance shows that certain genetic factors are transmitted according to the mode of transmission of the X chromosome, thus providing a direct correlation between the inheritance of a specific genetic trait with the inheritance of a specific chromosome.*

FURTHER READING

Morgan, T. H., 1910. "Sex-limited inheritance in Drosophila," reprinted from *Science* **32**:20 in *Classic papers in genetics*, 1959. Peters, J. A. (ed.). Englewood Cliffs, N. J.: Prentice-Hall, Inc.

PROBLEMS

The answers to the following problems are to be given separately in terms of the two sexes.

1. A cross was made between two wild-type *Drosophila.* Their progeny were 187 *ras* (raspberry eye color) males, 194 wild-type males, and 400 wild-type females. Is *ras* a sex-linked gene? Explain. What are the parental genotypes? What are the genotypes of the F_1 wild-type females and what is their ratio?

2. A cross was made between a female heterozygous for the recessive genes *ct* (*cut* wings) and *se* (*sepia* eye color) and a *se* male. Among their female progeny the phenotypes were 1/2 wild type and 1/2 sepia. Among their male progeny the phenotypes were 1/4 wild type, 1/4 cut, 1/4 sepia, and 1/4 cut and sepia. Are either of the genes sex-linked? What are the genotypes of the parents and their offspring?

3. If a woman having normal color vision has a color-blind father what is the probability that her sons will be color blind if she marries a man with normal color vision? What genotypes are possible among her male and female offspring? What is the probability of her having a color-blind child if she marries a color-blind man and is this probability different depending upon whether the child is a male or a female?

chapter 6

The Linkage and Recombination of Genes

Although it would be extremely difficult to single out any one area of investigation in genetics that has been the most significant or has made the greatest contribution to the field, investigations of linkage and recombination of genes certainly have been among the most instructive and provocative. From the early investigations of linkage and recombination in *Drosophila* by T. H. Morgan and his students have emerged one of the most important concepts in the field of genetics—the genetic map, which delineates the spatial relationships that are present between genes in the chromosomes. In this chapter, gene linkage, gene recombination, and the genetic map in nucleate organisms will be examined.

LINKAGE AND CROSSING OVER

We have seen that when two pairs of alleles enter into a cross, four different kinds of meiotic products are possible; that is, the cross of $a + \times + b$ gives the meiotic products $a +$, $+ b$, $+ +$, and $a\ b$. If a and b are in nonhomologous chromosomes, they assort independently. Thus, since they are *unlinked*, the four meiotic products are found in equal numbers. However, when the four meiotic products are unequal in number, independent assortment is, by definition, excluded and the gene loci involved are considered to be *linked*.

A departure from independent assortment was observed in 1906 by two British geneticists, W. Bateson and R. C. Punnett, in experiments with sweet peas. When plants that were *purple*-flowered with *long* pollen grains were crossed to *red*-flowered plants with *round* pollen grains, the F_1 resembled the first parent. Therefore, the genes for *purple* flower color and *long* pollen grain are dominant. When the F_1 was inbred, it was found that the F_2 ratio of phenotypes was not 9:3:3:1, but that there was a great excess of the two parental types *purple, long* and *red, round.* When the F_1 from a different cross *(purple, round* × *red, long)* was inbred, the F_2 again showed an excess of the two parental types. The genes for *purple* and *red*, as well as those for *long* and *round*, however, segregated in the expected 3:1 fashion in the F_2 generation, indicating that there was a pair of alleles for flower color and a pair for pollen-grain shape. The distinctive feature of these crosses lies, therefore, in the lack of independent assortment of the pair of color alleles with the pair of alleles for pollen-grain shape.

An explanation of these apparently anomolous results was provided in 1910 by T. H. Morgan. He postulated that pairs of alleles that do not assort independently are held together, or linked, because they are located in the *same chromosome* and that the recombinant types arise through some mechanism that breaks the linkage between them. Crossing over (see Chapter 2) was the mechanism proposed to explain the origin of the recombinant types. Crossing over will be discussed in this section in terms of data derived from tetrad analysis in a haploid organism and from strand analysis in a diploid.

In *Chlamydomonas* the two mutant genes *arg* and *pab* (a *pab* strain requires the vitamin *para*-aminobenzoic acid for growth) do not assort independently. When tetrads from a cross between the two were analyzed, the following classes and numbers of tetrads were obtained:

PD	NPD	T
arg +	*arg pab*	*arg* +
arg +	*arg pab*	*arg pab*
+ *pab*	+ +	+ *pab*
+ *pab*	+ +	+ +
119	1	71

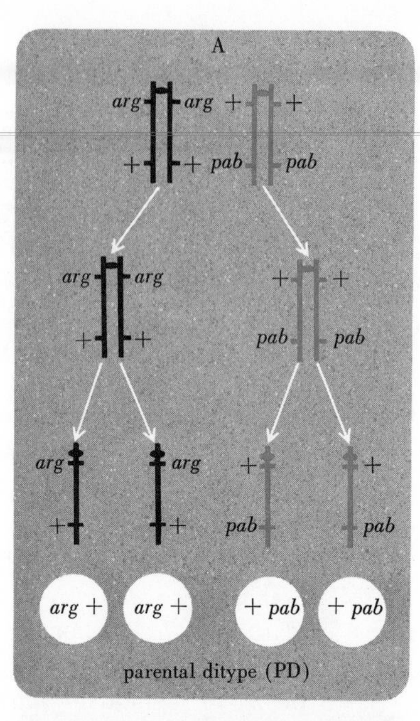

***Fig. 6-1** Origins of tetrads in a cross of* arg + × + pab *in* Chlamydomonas.

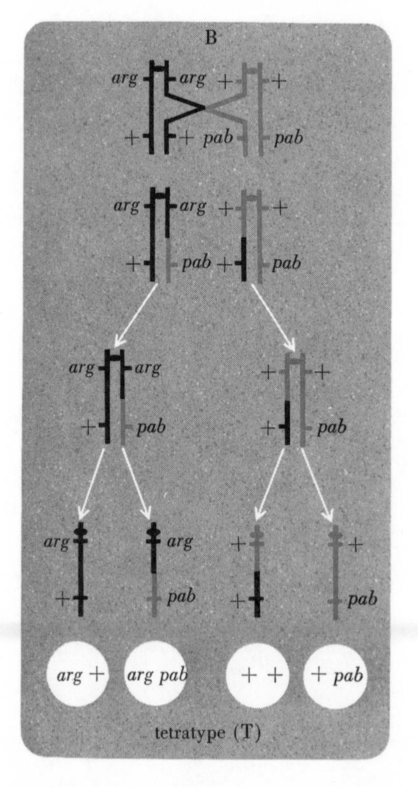

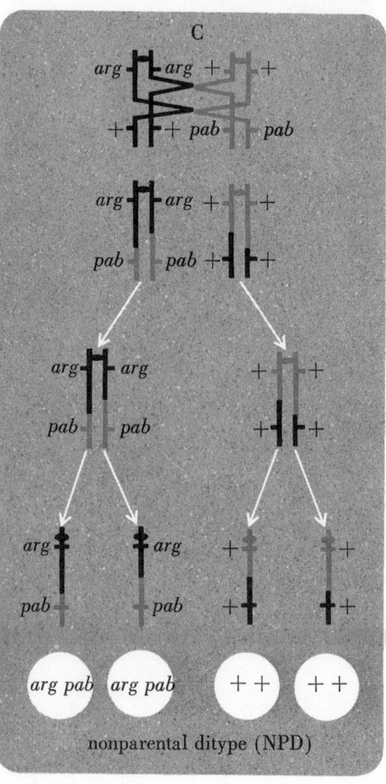

Quite clearly, the unequal ratio of PD to NPD tetrads is a major departure from independent assortment. This departure can be understood by determining the simplest origin of the two recombinant classes of tetrads (NPD and T). You will recall that when two genes are not linked (that is, when they are located in nonhomologous chromosomes), PD and NPD tetrads occur with equal frequency as the result of the independent alignment of the chromosomes at metaphase I of meiosis (Chapter 4). From the cross *arg* + × + *pab*, however, it is clear that there is an unequal ratio of PD to NPD tetrads, and by examining the genotypes you will find that there are more parental or nonrecombinant progeny than recombinant progeny. Let us consider the origin of the three classes of tetrads, assuming that the mutant genes *arg* and *pab* are linked.

The simplest origin of T tetrads is through a single exchange, or *single crossover*, between the *arg* and *pab* loci involving two of the four chromatids (Fig. 6-1B). All four products in an NPD tetrad are recombinant and, therefore, two crossovers involving all four chromatids must have occurred. The *double crossover* that yields an NPD tetrad is shown diagrammatically in Fig. 6-1C.

It is seen, therefore, that the PD tetrads arise when there is no crossing over between the *arg* and *pab* gene loci (Fig. 6-1A), whereas the T and NPD tetrads arise as the result of crossing over. Of the total tetrads, 62.3 percent are the PD or *noncrossover* class, 37.2 percent are the T class, and 0.5 percent are the NPD class. Thus, in this example, the tetrads that are the result of crossing over are less frequent than the PD or noncrossover tetrads. Furthermore, a T tetrad, formed by a single crossover, occurs more frequently than an NPD tetrad, which must arise through a double crossover. Two genes are said to assort independently when parental and recombinant types are equal in number and the PD:NPD ratio is one. *Two genes are defined as being linked when recombinant types are less frequent than the parental types.* In this case *the PD:NPD ratio is significantly greater than unity.* In terms of the event of crossing over, this statement implies that during meiosis two genes are held together more frequently than they are separated, and when they are separated, a single crossover is more likely to occur than a double crossover.

The percentage of recombination can be obtained from tetrads in the following way. Since each NPD tetrad has four recombinant strands, and each T only two, the total number of recombinant tetrads is equal to the sum of all NPD tetrads plus one half of the sum of all T tetrads. Therefore, the percent of recombination is equal to

$$\frac{\text{NPD} + \frac{1}{2}\,\text{T}}{\text{PD} + \text{NPD} + \text{T}}\,(100)$$

For the given data, the percentage of recombination between *arg* and *pab*, as determined from tetrad analysis, is

$$\frac{1 + \frac{1}{2}(71)}{191}(100) \quad \text{or} \quad \frac{36.5}{191}(100) = 19.1 \text{ percent}$$

The frequency of recombination between *arg* and *pab* may also be obtained by dividing the number of recombinant individuals or strands by the total number of individuals examined. In our example from *Chlamydomonas* the number of tetrads is 191. Therefore, the total number of strands is 4 × 191 or 764. The total number of recombinant strands is determined from the recombinant, or NPD and T, classes of tetrads. The single NPD tetrad has four recombinant strands, whereas half of the strands of the T tetrads are recombinant and half are parental. There are 71 T tetrads, and therefore 284 strands. Of these, half or 142 are recombinant strands. The total number of recombinant strands therefore is 4 + 142 or 146. The frequency of recombinant strands is 0.191 (that is, 146/764) or 19.1 percent.

The same result would have been obtained if individual meiotic products had been collected at random rather than from tetrads. This random collection is essentially what is done in strand analysis with diploid organisms. For example, in *Drosophila* the two mutant genes *glass* eye *(gl)* and *ebony* body color *(e)* do not assort independently. In homozygous flies the linkage between them is symbolized by a connecting line between *gl* and *e;* that is, $\frac{gl\ e}{gl\ e}$. Their wild-type alleles are thus given as $\frac{+\,+}{+\,+}$. When a cross of

$$glass\ ebony \times \text{wild type}\ \left(\frac{gl\ e}{gl\ e} \times \frac{+\,+}{+\,+}\right)$$

is made, the F_1 is wild in phenotype and has the genotype $\frac{gl\ e}{+\,+}$. A test cross of the F_1 females to $\frac{gl\ e}{gl\ e}$ males reveals the kind and frequency of gametes produced by this heterozygote. The data in Table 6-1 show that the $\frac{gl\ e}{+\,+}$ females produce four kinds of gametes with respect to the alleles under consideration. Unlike the instances of independent assortment seen in Chapter 4, however, the four kinds of gametes are not produced in a 1:1:1:1 ratio. The parental types (*gl e* and + +) represent 92 percent of all gametes produced, whereas there are only 8 percent of the recombinant or nonparental types (*gl* + and + *e*). Thus, *gl* and *e* and their wild-type alleles do not assort independently and are, therefore, linked. In other words, they remain linked to each other 92 percent of the time, and they recombine 8 percent of the time.

Table 6-1 **Linkage and Recombination between *gl* and *e* in *Drosophila***

$$\frac{gl\ e}{+\ +} \times \frac{gl\ e}{gl\ e}$$

	Number	Percent	
$\frac{gl\ e}{gl\ e}$	1031	43.4	
$\frac{+\ +}{gl\ e}$	1159	48.6	92.0
$\frac{gl\ +}{gl\ e}$	92	3.8	
$\frac{+\ e}{gl\ e}$	99	4.2	8.0

Linkage is also shown in the cross

$$glass \times ebony \left(\frac{gl\ +}{gl\ +} \times \frac{+\ e}{+\ e}\right)$$

Once again the F_1 is wild type, but the F_1 genotype is now symbolized $\frac{gl\ +}{+\ e}$. This genotype indicates that in these F_1 flies, *glass* and the wild-type allele of *ebony* were contributed by one parent; the other parent contributed *ebony* and the wild-type allele of *glass*. If these F_1 females are test crossed to $\frac{gl\ e}{gl\ e}$ males, the linkage between the *gl* and *e* loci is again approximately 92 percent, and the recombinant types occur approximately 8 percent of the time. This second cross shows that the manner in which the mutant loci enter the cross has no effect on the frequency of linkage and recombination.

The genes *gl* and *e* recombine about 8 percent of the time. Thus, in 16 of 100 meioses in a heterozygous female the linkage between the two genes is broken by crossing over, but in 84 meioses it is retained (Table 6-2). It should be pointed out that in male *Drosophila* there is no crossing over in any of the chromosomes, and thus linkage is complete. Complete linkage such as this is exceptional, and it is found only in a limited number of organisms.

LINKAGE GROUPS

Investigations with many different organisms have shown that each species is characterized by a limited number of groups of linked genes. These groups are referred to as *linkage groups*. Any genes that show linkage are in the same group, whereas those that assort independently are

Table 6-2 **The Relationship between Crossover Frequency and Recombination Frequency**

From 100 Meioses		*Percent Parental Strands from Each Meiotic Event*	*Percent Recombinant Strands from Each Meiotic Event*	*Number of Parental Strands Recovered*	*Number of Recombinant Strands Recovered*
gl *e* / + +	84	100	0	84	0
gl *e* / + + (X)	16	50	50	8	8
				Total 92	8

in different groups. Moreover, it is seen that the number of linkage groups possessed by a given organism corresponds to its number of chromosome pairs, and thus to its haploid chromosome number. The number of linkage groups in *Drosophila melanogaster* (the species of *Drosophila* most often referred to in this text) is four, and there are four pairs of chromosomes. On the other hand, another species, *Drosophila pseudoobscura*, has five pairs of chromosomes and also five linkage groups. Corn has ten linkage groups and ten chromosome pairs. This correspondence between the number of linkage groups and the number of chromosome pairs is further evidence in support of the chromosome theory, and it shows the close relationship between the meiotic behavior of chromosomes and the transmission of the genetic material.

GENETIC AND CYTOLOGICAL PROOF OF CROSSING OVER

The preceding discussion has assumed that recombination results from an exchange or a crossover between a pair of nonsister chromatids. Two different experiments, one with maize and the other with *Drosophila*, have shown that crossing over entails some form of physical exchange between chromatids. One of the most elegant proofs for this comes from an experiment with *Drosophila* reported in 1931 by Curt Stern.

The purpose of Stern's experiment was to demonstrate that flies having recombinant genotypes for a pair of linked genes were also recombinant for the pair of chromosomes in which the genes were located. To prove this point, Stern had to be able to recognize the results of crossing over both genetically and cytologically. This requirement was met by the use of two strains of *Drosophila*, each of which had special X chromosomes morphologically different from each other and from the normal X chromosome. One strain arose spontaneously, whereas the other was obtained by treating normal flies with x rays. X irradiation can cause chromosomes to break and, in certain instances after irradiation, unions can occur between nonhomologous chromosomes. In Stern's experiment, strain 1 possessed X chromosomes carrying a segment of the Y chromosome. In strain 2, each of the X chromosomes was broken into two equal segments, and one of the two segments was joined to the fourth chromosome. Strain 2 carried two mutant genes, the dominant *Bar* eye *(B)* and a recessive eye-color mutant known as *carnation (car)*. Strain 1 carried their wild-type alleles.

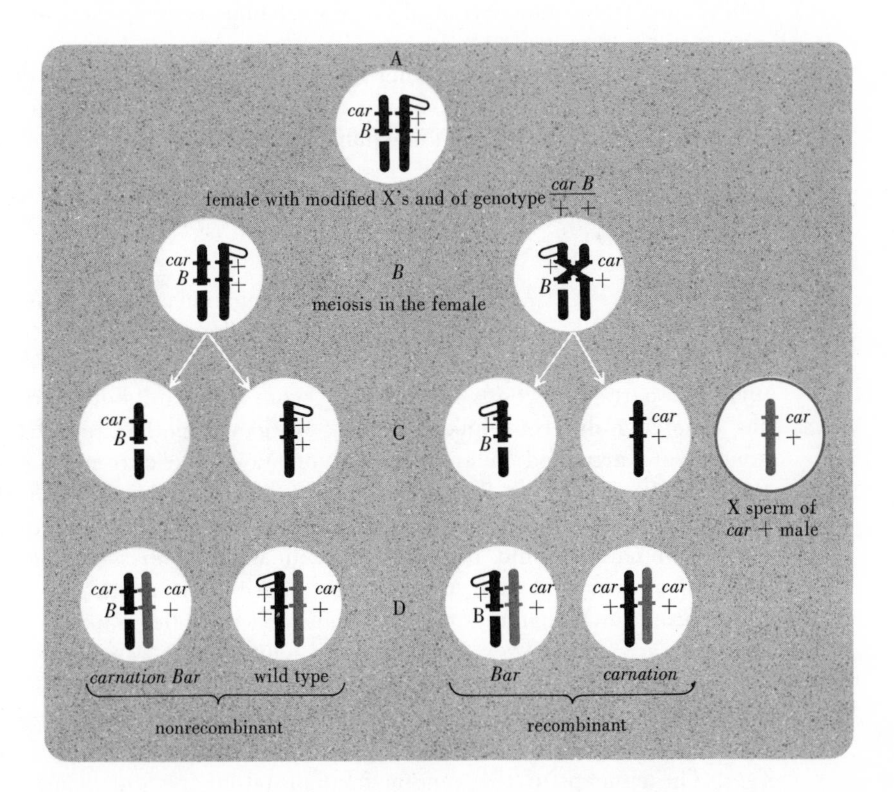

Fig. 6-2 *The results of Stern's experiment on the genetic and cytological proof of crossing over in* Drosophila *(see text for details). (After Altenburg.)*

The two strains were crossed to give females heterozygous for *car* and *B* and with X chromosomes that were morphologically different (Fig. 6-2A). Phenotypically these F_1 females had a wild-type eye color and an eye shape characteristic of heterozygous *Bar.* The F_1 females were crossed with *car* + males having morphologically normal X chromosomes. Stern then examined the female progeny of this cross and determined their phenotypes and the morphology of their X chromosomes. The results are shown in Fig. 6-2D. All of the females have at least one normal chromosome *(car +)* from their male parent. Half of the parental-type females have one of the special X chromosomes and half have the other. The recombinant females, on the other hand, have an X chromosome that is recombinant in morphology. One of these recombinant chromosomes is normal; the other carries the segment of the Y chromosome and is broken into two equal segments. Figure 6-2B shows the origin of these recombinant flies and their recombinant chromosomes. The results of Stern's experiment, which couple genetic and cytological evidence, support the contention that gene recombination results from some sort of physical exchange, or crossover, between chromatids. Similar experiments in maize by H. Creighton and B. McClintock have given further support. These experiments do not, however, provide any evidence for the mechanism whereby crossing over occurs. Several different mechanisms have been proposed and they will be discussed at the end of this chapter.

THE GENETIC MAP Early in the investigation of gene recombination in *Drosophila* it became evident that the frequency of recombination between linked genes varied, depending upon the particular genes in question. From the results of many crosses involving different linked genes, Morgan hypothesized that the genes were arranged in a linear fashion along the chromosome and that the frequency of recombination between them reflected their relative positions. A. H. Sturtevant then suggested that the frequency of recombination could be used as a measure of distance between pairs of genes, and that their linear distribution along the length of a chromosome could form the basis of a genetic map. Gene loci that give a relatively low frequency of recombination would be closer together than those that give a higher frequency of recombination. The scale for the map would be the percent recombination, and one unit on the map would be equivalent to 1 percent of recombination or crossing over. The concept of the genetic map proposed by Sturtevant was a brilliant one in that it provided a means for determining the spatial

Table 6-3 **Kinds of Tetrads Obtained from the Cross *arg* + + × + *pab thi* in *Chlamydomonas***

Tetrad					Number
(1)	*arg* + +	*arg* + +	+ *pab thi*	+ *pab thi*	14
(2)	*arg* + +	*arg pab thi*	+ + +	+ *pab thi*	54
(3)	*arg* + +	*arg* + *thi*	+ *pab* +	+ *pab thi*	103
(4)	*arg* + +	*arg pab* +	+ + *thi*	+ *pab thi*	2
(5)	*arg* + +	*arg pab thi*	+ + *thi*	+ *pab* +	6
(6)	*arg* + *thi*	*arg pab* +	+ + +	+ *pab thi*	3
(7)	*arg* + *thi*	*arg pab thi*	+ *pab* +	+ + +	6
(8)	*arg* + *thi*	*arg* + *thi*	+ *pab* +	+ *pab* +	2
(9)	*arg pab* +	*arg pab thi*	+ + +	+ + *thi*	1

relationships between genes, and thus their localization to specific sites in the chromosome.

The construction of genetic maps is a relatively simple but somewhat detailed procedure, and we will consider only two general examples, one from tetrad analysis and one from strand analysis. Let us take an example from tetrad analysis first. Given three linked loci, *a*, *b*, and *c*, what is their frequency of recombination, and thus the distance between them, and what is their sequence along the chromosome?

In *Chlamydomonas*, a cross made between a mutant strain that requires arginine *(arg)* and one that requires both *p*-aminobenzoic acid *(pab)* and thiamine *(thi)* yielded the different kinds of tetrads shown in Table 6-3. The frequency of recombination and the sequence of the three genes can be determined by classifying each tetrad as a PD, NPD, or T for *each pair* of loci: *arg-pab*, *arg-thi*, and *thi-pab*. For example, tetrad (3) in Table 6-3 is a PD with respect to *arg* and *pab* and a T with respect to *arg* and *thi* and with respect to *thi* and *pab*. Tetrad (7) is a T with respect to *arg* and *pab*, NPD with respect to *arg* and *thi*, and T with respect to *thi* and *pab*. After each tetrad has been classified in this manner, the total number of PD, NPD, and T tetrads for each pair of loci can be determined, as shown in Table 6-4. The recombination percentages calculated from these data by using the formula given on page 93 are, therefore,

$$\textit{arg-pab:} \quad \frac{1 + \frac{1}{2}(71)}{191}(100) = 19.1$$

$$\textit{arg-thi:} \quad \frac{8 + \frac{1}{2}(167)}{191}(100) = 48$$

$$\textit{thi-pab:} \quad \frac{2 + \frac{1}{2}(121)}{191}(100) = 32.8$$

The distance between *arg* and *pab* is defined by their frequency of recombination—that is, 19.1 *map units.* Similarly, the distance between *pab* and *thi* is 32.8 map units. The sequence of the three loci must be *arg pab thi*, for if it were any other sequence (*pab arg thi*, for example), the greatest percent recombination for a pair of loci would be found for *pab* and *thi* rather than for *arg* and *thi*. Additional crosses using other mutant strains can establish the details of the genetic map. Figure 6-3 shows a portion of the map of the *Chlamydomonas* chromosome carrying the *arg*, *pab*, and *thi* loci.

You will note that the two distances, *arg–pab* = 19.1 and *pab–thi* = 32.8, are not additive; their sum is 51.9 whereas the distance calculated by the formula on page 93 is 48. This discrepancy can be explained if one counts the *actual* number of crossovers between each pair of loci. When these values are obtained and divided by the total number of tetrads a crossover frequency is obtained for each pair. One half of this frequency is the recombination frequency; this value multiplied by 100 is equal to the map distance (see Table 6-2 describing the relationship between crossover frequency and recombination frequency). When this procedure is followed the map distances are *additive*, because *all* crossovers between each pair of genes are included in the calculation.

The origin of each class of tetrad is shown in Fig. 6-4. For example, since tetrads of class (1) in Table 6-3 are PD with respect to the three loci, their simplest origin is without crossing over. Tetrads of class (2),

Table 6-4 **Number of PD, NPD, and T Tetrads from Data Given in Table 6-3**

	arg-pab	*arg-thi*	*pab-thi*
PD	119	16	68
NPD	1	8	2
T	71	167	121
Total	191	191	191
Map distance	19.1	48.0	32.8

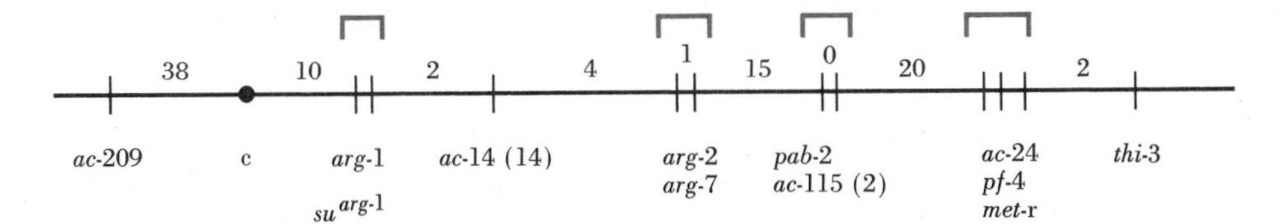

Fig. 6-3 *The genetic map of linkage group I of* Chlamydomonas reinhardi. *The symbols* c, ac, arg, su^{arg} *-1*, pab, pf, met-r, *and* thi *refer respectively to the centromere, acetate dependence, arginine dependence, suppressor of* arg, p-*aminobenzoic acid dependence, paralyzed flagella, methionine sulfoximine resistance, and thiamin dependence. The numbers above the line refer to map distances. A bracket above the line indicates either an unknown or uncertain map distance. A number in parentheses adjacent to a mutant symbol indicates the number of known alleles at the locus.*

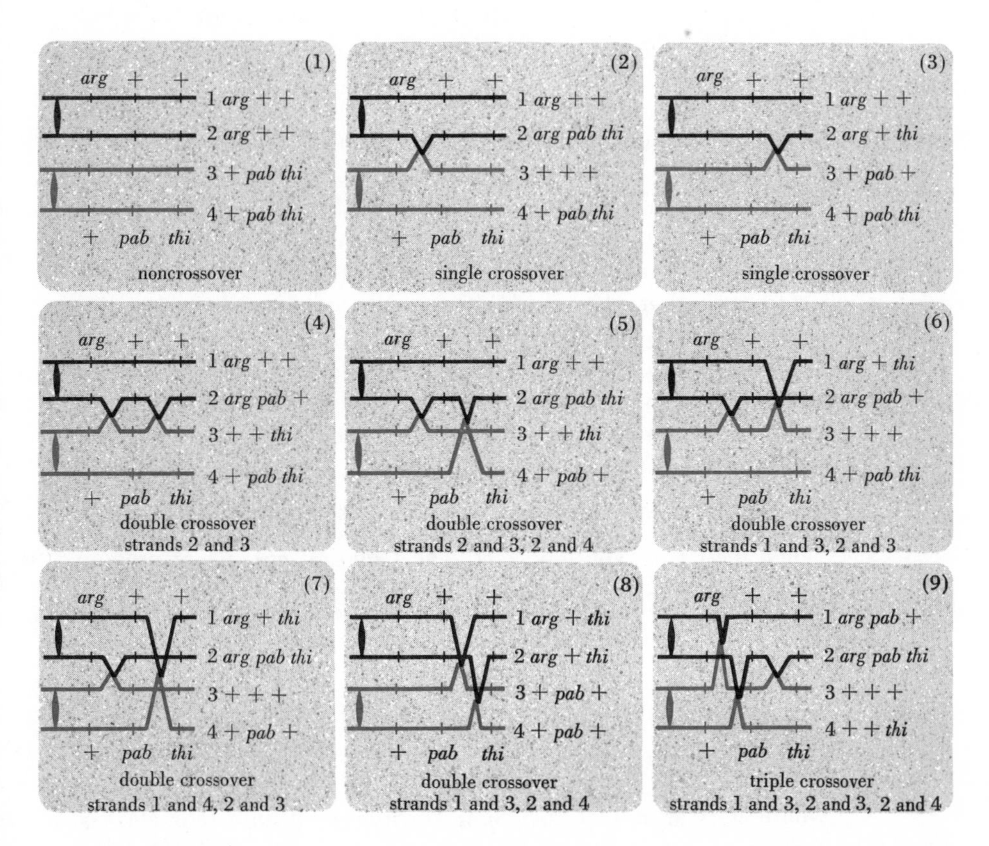

Fig. 6-4 *Origin of tetrads in a cross of* Chlamydomonas; arg + + × + pab

however, arise most simply by way of a *single crossover* in the region between the loci *arg* and *pab.* Tetrads of class (3) also arise by a single crossover, but in this case it is in the region between *pab* and *thi.* The remaining tetrads all entail the occurrence of at least two crossovers. The origin of all of the tetrads gives information that can be used to confirm the sequence determined earlier. Any sequence other than *arg pab thi* would shift one of the rarer, double-crossover class of tetrads into the more frequent, single-crossover class. For example, if the sequence were *arg thi pab*, then tetrads of class (4) would arise by a single crossover, whereas those of class (5) would require a two-strand double crossover. In addition to the nine classes of tetrads shown in Fig. 6-4, several others are possible, each reflecting other patterns of crossing over. They are not shown here since they were not found in this particular experiment.

By determining the crossover frequency it is found that there are 73 crossovers between *arg* and *pab*, 125 between *pab* and *thi* and 198 between *arg* and *thi.* The respective crossover frequencies are therefore 0.382, 0.654, and 1.04. Accordingly, the map distances are 19.1, 32.7, and 52.

In *Drosophila melanogaster*, each of the four pairs of chromosomes is well mapped (Fig. 6-5). Many different mutant strains are available and, since the early experiments of Morgan and his students, investigators have undertaken to localize each of the mutant genes. The mapping procedure is exemplified here by utilizing data obtained from crosses involving three different sex-linked genes. This procedure applies to the other chromosomes as well, and in general to other organisms with a diploid life cycle.

A typical cross is one involving three mutant strains, such as *yellow* body *(y)*, a rough eye known as *echinus (ec)*, and *cut* wings *(ct).* When females of a strain homozygous for *y ec ct* are crossed with wild-type males, the F_1 females are phenotypically wild type and heterozygous for all three loci $\left(\frac{y\ ec\ ct}{+\ +\ +}\right)$. The males are phenotypically *y ec ct* and hence their genotype is $\underrightarrow{y\ ec\ ct}$. The kind and number of progeny produced by the F_1 females can be determined by test crossing them to $\underrightarrow{y\ ec\ ct}$ males. We can predict that these females will produce eight different types of gametes: *y ec ct*, + + +, *y* + +, + *ec ct*, *y ec* +, + + *ct*, *y* + *ct*, and + *ec* +. We cannot, however, predict the number of each, since the gene loci in question are linked and therefore do not assort independently of each other. The results of the test cross are in Table 6-5, where it is seen that the eight genotypes fall into four classes. The first, class A, consists of two kinds of flies: *y ec ct* and + + +. They represent the parental or nonrecombinant class. Note that they form

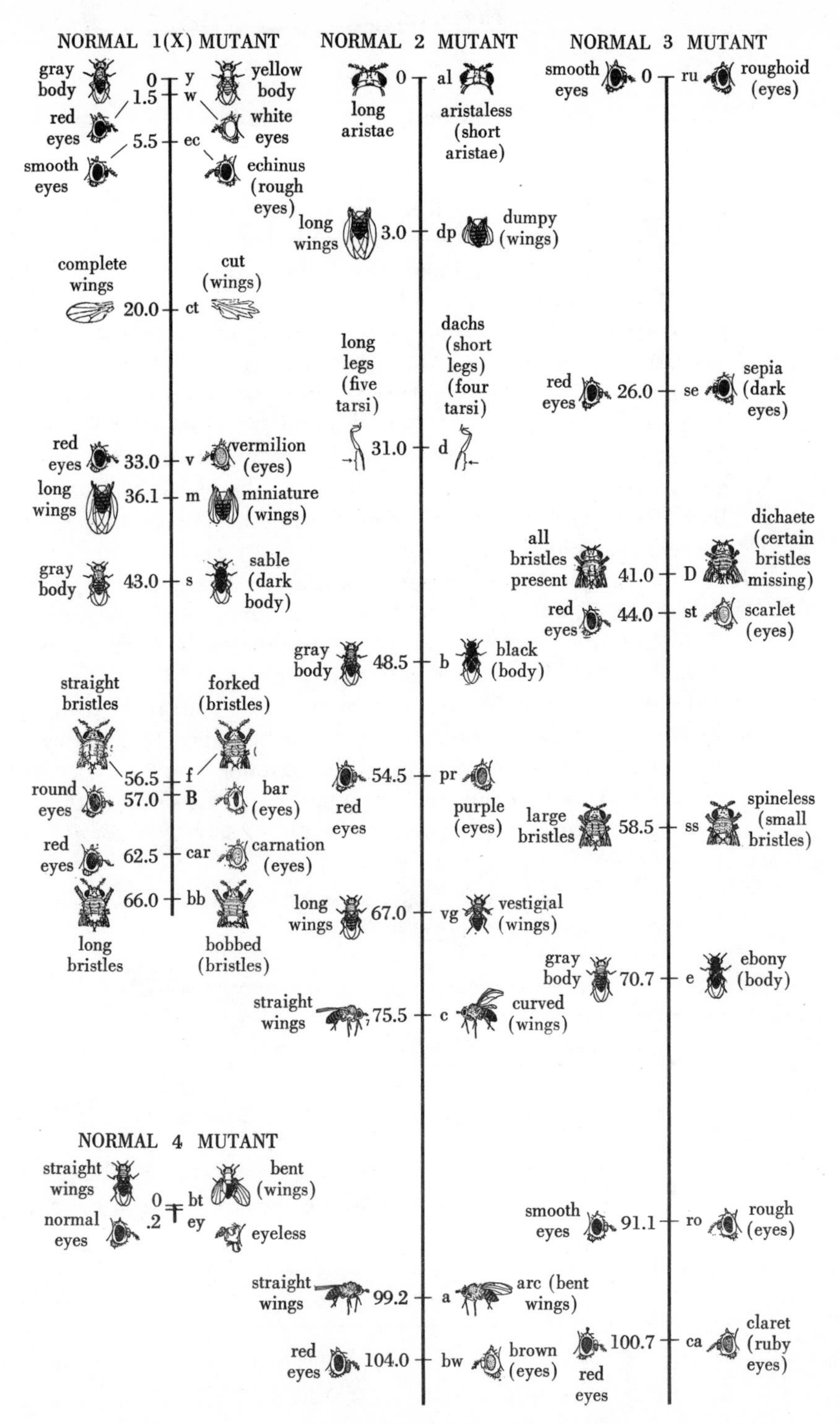

Fig. 6-5 *A portion of the genetic map of* Drosophila melanogaster.

the largest class of progeny. A second and smaller class (class B) consists of two kinds of flies with the two *complementary* genotypes *y* + + and + *ec ct*. This is a recombinant class. A second recombinant class (class C), also less frequent than the parental class, consists of flies that are *y ec* + and + + *ct*. A third and rare recombinant class (class D) is comprised of *y* + *ct* and + *ec* + flies.

The frequency of recombination between the three gene loci must be ascertained by determining the percent of recombination between any pair of gene loci. For instance, there are 154 flies that are recombinant for *y* and *ec*. These include the flies belonging to classes B and D in Table 6-5. Recombination between *y* and *ec* is accordingly 5.4 percent. The 585 recombinants for the gene loci *ec* and *ct* are found in classes C and D. Recombination between them is, therefore, 20.3 percent. The flies of classes B and C are recombinants between *y* and *ct*, and their percent recombination is 24.9.

With these data it is now possible to determine the sequence of the loci *y*, *ec*, and *ct*. Three different sequences are possible: *y ec ct*, *y ct ec*, and *ec y ct*. The distances, expressed in terms of percent of recombination, have been determined as $y - ec = 5.4$, $y - ct = 24.9$, and $ec - ct = 20.3$. Clearly, the sequence which fits these data is *y ec ct*, for the sum of *y*–*ec* and *ec*–*ct* distances is equal to 25.4, which is in good agreement with 24.9, the value obtained for the *y*–*ct* distance.

A way of verifying the sequence is by considering the possible crossover or recombinant classes. We have seen that linkage between

Table 6-5 **Linkage and Recombination between *y*, *ec*, and *ct* in *Drosophila***

$\frac{y\ ec\ ct}{+\ +\ +} \times \frac{y\ ec\ ct}{\rightarrow}$

	Phenotypes	Number	
A	*y ec ct* + + +	1071 1080	noncrossover
B	*y* + + + *ec ct*	78 66	single crossover between *y* and *ec*
C	*y ec* + + + *ct*	282 293	single crossover between *ec* and *ct*
D	*y* + *ct* + *ec* +	4 6	double crossover: one between *y* and *ec* and one between *ec* and *ct*
		2880	

genes is established when the recombinant classes occur less frequently than the parental class. In the case of these three linked loci, the two single-crossover classes are rarer than the parental class, and the double-crossover class is even rarer. Therefore, in the example under consideration here, a sequence other than *y ec ct* would shift the double-crossover class into one or the other single-crossover classes. For example, if the sequence were *ec ct y*, a test cross of $\frac{ec\ ct\ y}{+\ +\ +}$ females would give *ec* + + and + *ct y* flies as one of the single-crossover classes, and a double-crossover class consisting of *ec* + *y* and +*ct* + flies. The data presented in Table 6-5 show that the former class (class D) represent about 0.3 percent of the progeny and the latter class (class C) comprises 20 percent. We must conclude that the crossover classes have been improperly assigned; therefore, the order cannot be *ec ct y*. Similar reasoning will show that the order cannot be *ec y ct*, and only the order *y ec ct* will fit the observed data.

Genetic maps have been constructed for many different organisms. In each instance it has been possible to show that the genes are arranged in a linear fashion in the chromosomes, and that the distances between groups of linked genes are essentially additive, just as in the examples given here. The accuracy of a genetic map increases as additional mutant genes become available for crosses. The frequency of multiple crossovers (double crossovers, triple crossovers, etc.) will affect the accuracy of the map, so that in mapping experiments an effort is usually made to use gene loci that are relatively close together.

INTERFERENCE AND COINCIDENCE

In the foregoing example from *Drosophila*, the double-crossover class comprised 0.34 percent of the progeny. Consider that this class arises through the simultaneous and independent occurrence of two single crossovers, one between *y* and *ec* and another between *ec* and *ct*. The *expected* frequency of double crossovers should be the product of the frequency of recombination between *y* and *ec* times the frequency of recombination between *ec* and *ct* since the probability of the simultaneous occurrence of two independent events is the product of their separate probabilities. The frequency of recombination between *y* and *ec* was 0.054 and between *ec* and *ct* was 0.203. The expected frequency of double crossovers is (0.054) (0.203) = 0.011 or 1.1 percent. The observed value was only 0.34 percent. The discrepancy between the expected and observed figures indicates that the two crossover events are not independent of each other. It appears as if the occurrence of a crossover at one position in a chromosome decreases the probability of a crossover occurring at another position. This lack of independence is called *interference*, and while it is a general phenomenon in linkage, little is known about its nature. Interference is usually expressed as *coincidence*. Coincidence is the ratio of the observed double crossovers to the expected double crossovers. When coincidence equals one, there is no interference; when it is zero, interference is positive and complete. Values between zero and one are indicative of partial interference. In the example given here, coincidence is $0.34/1.10 = 0.31$; that is, only 31 percent of the expected double crossovers were found. As a general rule, interference increases as distances between gene loci decrease.

THE GENE AND RECOMBINATION

From what you have learned of segregation and recombination, you might conclude that a gene is an indivisible unit of genetic transmission. In fact, a gene is not an indivisible unit of heredity; it is composed of a sequence of nucleotides that determines the sequence of amino acids in a protein. A mutation can occur anywhere in this chain of nucleotides, giving rise to different proteins and hence to different modifications of the same basic phenotype. Recombination can occur between two mutant alleles that affect the same function just as recombination can occur between two mutant alleles that affect different functions.

As an example, let us consider a mutant form of *Drosophila* known as *lozenge (lz)*. In phenotype, lozenge flies have an eye that is rougher in texture than that of wild-type flies. Three different alleles, lz^{g}, lz^{BS}, and lz^{46} produce *lozenge* phenotypes that are quite similar, yet distinguishable on the basis of different modifications in the texture and shape of the eye. The *lz* alleles are recessive and in the X chromosome. Early experiments localized them at a distance of 27.7 map units from one end of the chromosome, and crosses between different *lz* mutants showed no recombination. It was thus assumed that the three were alleles at the same locus.

Later experiments by the geneticists M. and K. Green showed that this was not the case. The cross $\left(\frac{lz^{BS}}{lz^{46}} \times \frac{lz^{BS}}{\rightarrow}\right)$, for example, usually gives male progeny that are 50 percent lz^{BS} and 50 percent lz^{46}. However, in a sample of male progeny as large as 20,000 there will be from five to ten *wild-type* males. These wild-type males were shown to have arisen by crossing over *between* the *lz* alleles, and the reciprocal cross-over type $lz^{BS}lz^{46}$ was also recovered. Thus the three *lz* alleles do not occupy the same locus. Recombination can occur between

lz^{BS}, lz^{46}, and lz^{g}, and the three are located in that order within a segment of the X chromosome 0.2 map unit in length.

The *lz* alleles might be considered to be different genes from the point of view of recombination. However, from the functional view they are not. There is one *lz* gene occupying a chromosome segment 0.2 map unit in length. Within this gene there are at least three sites that can mutate independently of each other and that recombine at low frequency. Therefore, it is more appropriate to define a gene as a unit of function than as a unit of recombination.

If a gene determines a function, then what is the phenotype of an organism heterozygous for two different mutations *within* the same gene? This question can be answered by examining females heterozygous for different *lz* alleles. It is found that a female heterozygous for only one *lz* allele has the wild phenotype but that females heterozygous for two different *lz* alleles are either wild or mutant in phenotype depending on the position of the alleles with respect to one another. Two different positions are possible. Either both alleles are on the same chromosome *(cis)*, or they are on different chromosomes *(trans)*. A female that is a *cis* heterozygote $\frac{lz^{BS}\ lz^{46}}{+\ \ +}$ has the wild phenotype, but a *trans* heterozygote $\frac{lz^{BS}\ \ +}{+\ \ lz^{46}}$ is mutant in phenotype. The *cis* heterozygote shows that neither of the *lz* alleles is dominant and that the necessary genetic information for the wild phenotype is obtained from the wild–type chromosome. However, since the *trans* heterozygote is mutant, it must mean that neither chromosome has the information required for the wild phenotype. The lz^{BS} + segment of the chromosome is defective for *part* of the *lz* locus and the + lz^{46} chromosome is defective for a *different* part. This *cis-trans* effect is shown only between alleles that are very closely linked and whose phenotypic effect—and presumably their functions—are similar.

Closely-linked genes with *different* functions do not exhibit a *cis-trans* effect. Two adenine-requiring mutants of the fungus *Aspergillus* may serve as an example. The alleles, *ad*-8 and *ad*-16, are closely linked to the recessive yellow *(y)* allele that determines conida color. Recombination tests have shown that the three loci are arranged in the order *y ad*-16 *ad*-8; the distance between *y* and *ad*-16 is 0.05 map unit and the distance between *ad*-16 and *ad*-8 is 0.1 unit. Diploid forms of *Aspergillus* can be obtained. The heterozygotes $\frac{y\ ad\text{-}16}{+\ \ +}$ and $\frac{y\ \ +}{+\ ad\text{-}16}$ are both wild in phenotype; no *cis-trans* effect is shown. However, the heterozygotes $\frac{ad\text{-}16\ \ +}{+\ \ ad\text{-}8}$ and $\frac{ad\text{-}16\ ad\text{-}8}{+\ \ +}$ show the *cis-trans* effect, for the *trans* heterozygote is mutant and the *cis* heterozygote is wild type. The *cis-trans* effect, therefore, is shown only between the functionally related *ad* mutants, and not between the *ad* mutant and the *y* locus that affects the pigment of the conidia.

The *cis-trans* effect is explained in terms of the phenomenon called *complementation.* The two adenine loci of *Aspergillus* are said to be *noncomplementary* and functionally alike since the *trans* heterozygote is mutant and will not grow on minimal medium. The functional distinction between them can be demonstrated easily by forming *heterokaryons* between the *ad-16* and *ad-8* strains. Heterokaryons can be formed by a number of different fungi. They consist of cells of the same mating type carrying a mixture of two genetically different kinds of nuclei within the same cytoplasm. When a heterokaryon between two biochemical mutants is formed it will either grow or fail to grow on minimal medium. The *ad-8 ad-16* heterokaryon fails to grow on minimal medium.

A gene can be defined from the functional point of view by a *cis-trans* test. The fact that *ad-8* and *ad-16* are noncomplementary implies identity of function; that is, the same function has been lost by both *ad-8* and *ad-16.* The important point is that the two mutants, which arose by independent events of mutation and which are spatially separated in the chromosome, are functionally alike. Both the *cis-trans* test in the diploid, and the heterokaryon test for complementation show that no genetic information is available for the production of the wild phenotype. From the functional point of view, therefore, *ad-8* and *ad-16* are in the same gene even though they map to different sites within this gene.

A gene, defined from the functional point of view by a *cis-trans* test, is called a *cistron.* The cistron in turn can be divisible by recombination, and it possesses many sites at which mutation can occur. It will become increasingly clear that a protein is the ultimate product of the functional genetic unit, and, as you will learn in Chapter 10, a given unit specifies the nucleotide sequence of *messenger* RNA which in turn specifies the amino acid sequence of a given protein.

POSSIBLE MECHANISMS FOR GENE RECOMBINATION

Among the hypotheses put forward to explain gene recombination in viruses and both nucleate and nonnucleate organisms, two have received considerable attention. They are the *chromosome breakage and joining hypothesis* and the *copy-choice hypothesis.* According to the breakage hypothesis, a break can occur at the same level in a pair of nonsister chromatids and, following the break, the nonsisters join together. The result is two reciprocally

Fig. 6-6 Breakage model for gene recombination: (A) relational coiling of chromatids; (B) the breakage and reunion of nonsister chromatids.

recombinant chromosomes (Fig. 6-6). The copy-choice or Belling hypothesis (named after the cytologist J. Belling who first proposed it) assumes that recombination occurs during the process of chromosome duplication and thus during DNA replication. As a pair of homologues duplicates, the process of duplication or "copying" can proceed partially along one homologue and partially along the other. If this should occur, a crossover would be the result (Fig. 6-7).

There is, however, at least one major difficulty with this hypothesis. You will note that according to this hypothesis only the two newly formed strands can participate in a copy-choice type of recombination. The two original strands remain intact. In fact, tetrad analysis has demonstrated that all four chromatids can participate in crossing over as shown by the fact that multiple-strand double crossovers occur (Fig. 6-4). Furthermore, crossing over occurs at a four-strand stage and it is reciprocal. The fact that *all four chromatids* participate in crossing over cannot be reconciled with the copy-choice hypothesis unless the assumption is made that a mechanism exists allowing crossing over to occur between sister strands.

There is evidence that recombination in phage can occur by the breakage and joining of double-stranded DNA molecules, and evidence is accumulating to indicate that this kind of recombination occurs in bacteria as well. Although there is, as yet, no direct evidence to show that recombination can occur by the breakage and joining of double-stranded DNA molecules in nucleate organisms, the properties of

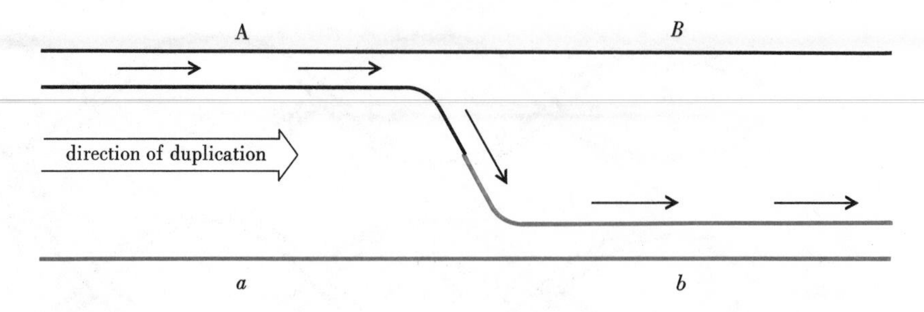

Fig. 6-7 *The copy-choice model for gene recombination (explanation in text).*

recombination as inferred from tetrad analysis are those to be expected from chromosome breakage and joining rather than from copy choice. Some fraction of recombination in all organisms may possibly occur by a copy-choice mechanism, but there is no evidence for it at a molecular level.

PROBLEMS

1. An analysis of tetrads in a cross of *Chlamydomonas* involving the loci *pf (paralyzed flagella)* and *nic* (requirement for the vitamin nicotinamide) gave the following results: PD = 70, NPD = 2, and T = 28. Since the PD to NPD ratio does not equal one, *pf* and *nic* must be linked. Using the methods described on page 100 calculate the frequency of recombination between these two loci.

2. In *Chlamydomonas* the genes designated here as *a*, *b*, and *c* are linked. The following are the tetrads recovered from a cross involving these three loci.

	(1)	(2)	(3)	(4)	(5)	(6)	(7)
	a b c	*a b c*	*a b c*	*a b c*	*a* + +	*a b* +	*a b c*
	a + +	*a* + *c*	*a b c*	*a* + +	*a b* +	*a* + *c*	*a b* +
	+ *b* +	+ *b* +	+ + +	+ *b c*	+ + *c*	+ *b c*	+ + *c*
	+ + *c*	+ + +	+ + +	+ + +	+ *b c*	+ + +	+ + +
Number of tetrads	10	10	440	160	10	10	360

(a) What are the genotypes of the parents?
(b) Determine the map distances between *a*, *b*, and *c*.
(c) Determine the simplest origin of each of the seven classes of tetrads.
(d) What is the coefficient of coincidence?

3. Analysis of ordered tetrads in *Neurospora* from a cross $a\ b\ c \times +++$ gave the following tetrad types:

	(1)	(2)	(3)	(4)	(5)	(6)
	$a\ b\ c$	$a\ b\ +$	$a\ b\ c$	$a\ b\ +$	$a\ b\ c$	$a\ b\ +$
	$a\ b\ c$	$a\ b\ +$	$+\ b\ c$	$+\ b\ +$	$a\ +\ c$	$a\ +\ +$
	$+++$	$+\ +\ c$	$a\ +\ +$	$a\ +\ c$	$+\ b\ +$	$+\ b\ c$
	$+++$	$+\ +\ c$	$+++$	$+\ +\ c$	$+++$	$+\ +\ c$
Number of tetrads	300	300	100	100	100	100

(a) Are any of the genes linked? If so, which ones?
(b) If any of the genes are linked, how far apart are they?
(c) Calculate the distance between each gene and its centromere.
(d) Construct a genetic map from the analysis of the tetrad data.

4. In *Drosophila* three linked genes, *a*, *b*, and *c* gave the following results when *a c* females were crossed to *b* males and the F_1 inbred.

F_1: The phenotypes were + females and *a c* males

F_2:

Phenotypes	Number of Females	Number of Males
a	49	2
b	0	428
c	49	48
+	451	23
a b	0	47
a c	451	428
b c	0	1
a b c	0	23
	1000	1000

(a) Are the three genes sex-linked or autosomal?
(b) What are the parental and F_1 genotypes?
(c) What is the percent recombination between the three genes?
(d) What is the sequence of the three genes in their chromosome?
(e) What is the coefficient of coincidence and is there any interference?

5. In *Drosophila* the sex-linked genes *cut* (*ct*), *lozenge eyes* (*lz*), and *forked* bristles (*f*) are the following map distances apart: *cut* to *lozenge* 7.7 units, and *lozenge* to *forked* 29.0 units. Assuming that there is no interference, what are the expected numbers of genotypes out of 1000 flies recovered from the cross $\frac{ct\ lz\ f}{+++} \times \frac{ct\ lz\ f}{\longrightarrow}$? What effect will a coefficient of coincidence of 0.5 have on your answer?

6. Heterokaryons were made between seven different mutant strains of *Neurospora*, each of which required arginine for growth. They were tested for growth on minimal medium. The following results were obtained where + means growth and − means no growth on minimal medium.

Mutant Strain

Mutant Strain	1	2	3	4	5	6	7
1		+	−	+	+	+	−
2			+	−	+	+	−
3				+	+	+	−
4					+	+	−
5						−	−
6							−

(a) How many cistrons are there?
(b) Which mutations are in the same cistron?
(c) How do you explain the behavior of mutant strain number seven? How would you test your explanation?

chapter 7

The Cytoplasm in Heredity

The preceding chapters have emphasized that DNA is the genetic material and that it is transmitted in a highly predictable manner. In nucleate organisms the pattern of transmission is correlated with the behavior of the chromosomes during meiosis and mitosis, and thus with the genetic material in the nucleus rather than with some other components of the cytoplasm. Although we have seen ample evidence that this statement is true, the expression of certain hereditary traits can be influenced by the maternal genotype independently of the genotype of the offspring. Other hereditary traits can be attributed to the action of infectious, DNA-containing particles whose reproduction and action may or may not be dependent upon nuclear

DNA, and still others can be attributed to DNA that is found in organelles such as chloroplasts and mitochondria.

THE MATERNAL EFFECT

In the flower moth *Ephestia* the wild type has a pigmented larval skin and the adult has dark brown eyes. The pigmentation results from the presence of a substance called *kynurenine* whose synthesis is controlled by a gene *A*. In mutant strains *(a/a)* that lack kynurenine, the larvae are pigmentless and the adults have red rather than dark brown eyes. Crosses between pigmented and nonpigmented animals *(A/A* × *a/a)* result in an F_1 that is pigmented regardless of which parent is *A/A*. When the heterozygous F_1 progeny are test crossed, however, the sex of the pigmented parent appears to affect the phenotype of the F_2. In the cross of *A/a* males and *a/a* females, half of the larvae are pigmented and as adults have dark brown eyes. However, when the reciprocal cross is made, *A/a* females × *a/a* males, *all* of the larvae are pigmented, but among the adults half have dark brown eyes and half have red eyes. These results cannot be explained on the basis of sex-linked inheritance, since, for one, the F_1 is identical regardless of reciprocal crosses. When the F_1 is test crossed, the results obtained also differ from sex linkage. Determine for yourself that sex-linked inheritance cannot be invoked to explain the results.

When the female has the genotype *A/a* and is pigmented, she transmits this phenotype to *all* of her offspring regardless of their own genotype. She transmits, via the egg, sufficient kynurenine that the genotypically *a/a* progeny are pigmented, even though they lack the genetic information for synthesizing the pigment. However, as these animals develop there is an increasing dilution of the pigment; therefore, the larval color is lost and, ultimately, the adult has no brown eye pigment. When the pigmented parent in the test cross is a male, the results are an immediately apparent 1:1 segregation for the pigment trait. There is thus no carry-over of kynurenine in the sperm. This example of *maternal influence* or *maternal effect* is explained by the fact that the egg, with its large cytoplasmic content, contains products of the action of the maternal genotype, which in turn affect the phenotype of the progeny independently of their own genotypes. The effect may be transitory, as in the case of *Ephestia*, or it may be a permanent one that lasts throughout the life of the organism.

A permanent maternal effect is observed in the direction of the coiling of the shell of the snail *Limnaea*. The direction is determined at the time of the first cleavage division of the egg, and it depends solely upon the genotype of the female parent. Crosses between snails with a left-handed or sinistral coiling and those with a right-handed or dextral coiling are shown in Fig. 7-1. Note that the maternal genotype in each case determines the phenotype of its offspring, regardless of whether the offspring are genotypically dextral or sinistral. For both examples given here, *Ephestia* and *Limnaea*, the phenotypic expression of a single gene trait is affected by the genetic constitution of the female parent. Maternal effects such as these may play an important role in morphogenesis, since the early stages of cellular differentiation will take place within a cytoplasm whose contents have been elaborated by the action of the female's genotype. Further discussion of this important problem will be found in another text in this series, *Interacting Systems in Development* by James D. Ebert.

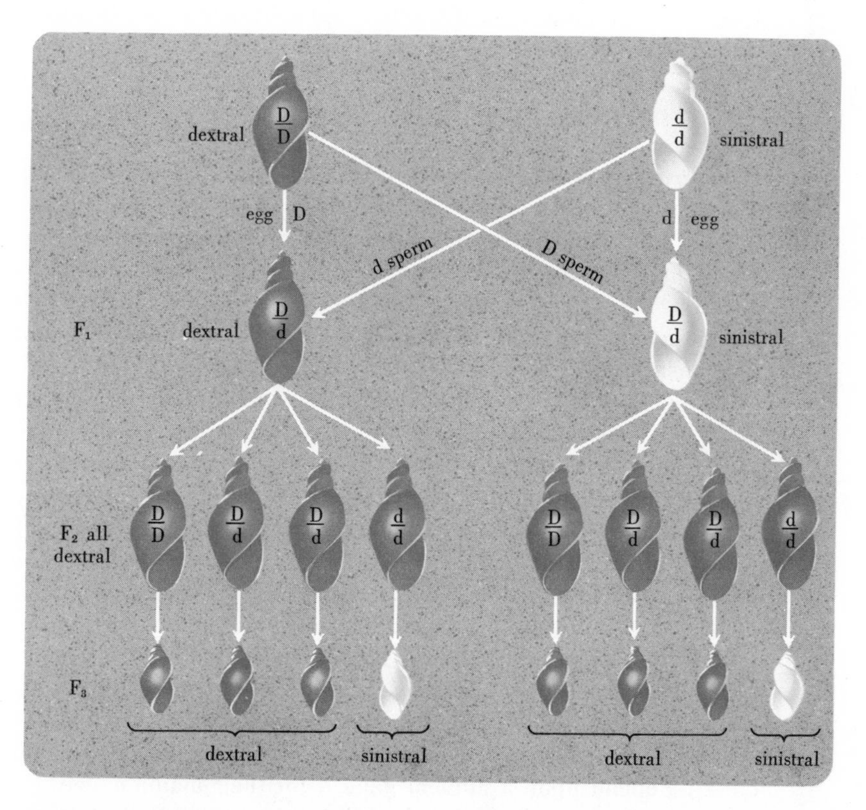

Fig. 7-1 *The inheritance of coiling direction in the shells of the snail* Limnaea. *(After Sinnot,* et al., Principles of Genetics, *5th ed. New York: McGraw-Hill, 1958.)*

INFECTIOUS HEREDITY In contrast to the maternal effect, which ultimately depends upon the action of maternal genes, there are forms of inheritance attributable to the action of cytoplasmic particles which are self-duplicating and transmitted from one cell generation to the next. Such cytoplasmic particles are, therefore, endowed with genetic continuity. Infectious heredity is one form of this sort of inheritance, and two examples will be considered here.

In *Drosophila*, sensitivity to carbon dioxide appears to be attributable to the presence of self-duplicating cytoplasmic particles. Most *Drosophila* are relatively resistant to carbon dioxide, and though the gas acts as an anesthetic, the flies recover once they are placed in air. The CO_2-sensitive animals, however, are killed by relatively low concentrations of the gas. There is a strain of *Drosophila* that breeds true for the character of CO_2 sensitivity, and when CO_2-sensitive females are crossed to wild-type males their offspring are almost entirely sensitive. When normal females are crossed to CO_2-sensitive males, however, CO_2-sensitive progeny are rarely obtained. CO_2-sensitive females obtained from either of these crosses will transmit the character to their offspring.

The marked difference in reciprocal crosses suggests that CO_2 sensitivity does not follow a classical form of inheritance. In an extensive series of crosses, CO_2-sensitive females were crossed to normal males; the female offspring from this cross again were crossed to normal males, and so on for a sufficient number of generations to replace all of the maternal chromosomes with those from the paternal parents. Progeny in all of the back crosses were found to be CO_2 sensitive, indicating that the trait is independent of the chromosomes, and must be transmitted in the egg cytoplasm. Further investigation showed that viruslike particles called *sigma* could be isolated from the CO_2-sensitive flies. The *sigma* particles have infectious properties, because when the ovaries of CO_2-sensitive animals are transplanted into normal females, these females become CO_2 sensitive.

T. M. Sonneborn has made extensive studies of a form of extranuclear inheritance in the ciliate protozoan, *Paramecium aurelia*. Certain strains of this organism, called "killers," release particles from their cytoplasm into the surrounding medium, and the particles contain a substance that brings about the death of "sensitive" cells. These particles, called *kappa* particles, contain DNA and are 0.2 to 0.8 μ in diameter. Thus, they are smaller than some bacteria yet larger than the viruses. Unlike the *sigma* particles of *Drosophila*, the *kappa* particles clearly depend upon a nuclear gene *K* for their maintenance.

In order to understand the inheritance of the killer character, it is necessary to examine the life cycle of *Paramecium aurelia*. *P. aurelia* is diploid and contains a macronucleus and two micronuclei. Conjugation occurs between cells of opposite mating type in the manner shown in Fig. 7-2, where only the micronuclei are shown. The micronuclei of conjugating cells undergo meiosis, after which all but one of the meiotic products disintegrate in each conjugant. The remaining haploid nucleus undergoes a single mitotic division and each of the conjugating cells donates one of these haploid nuclei to its mate. The two gamete nuclei then fuse to form a new diploid micronucleus. The conjugants separate and then reproduce asexually by fission to form clones. Under certain circumstances the process of conjugation is prolonged and then cytoplasm is exchanged as well as the nuclei. *P. aurelia* can also undergo self-fertilization by a process known as *autogamy*, in which any genotype produces only homozygous clones (Fig. 7-3). Thus, following autogamy the heterozygote *Aa* will produce either of two homozygous clones, *AA* or *aa*.

Returning to the killer character once again, its dependence on a nuclear gene *K* can be demonstrated when, by special techniques, killer and sensitive animals are crossed. Such a cross leads to the pro-

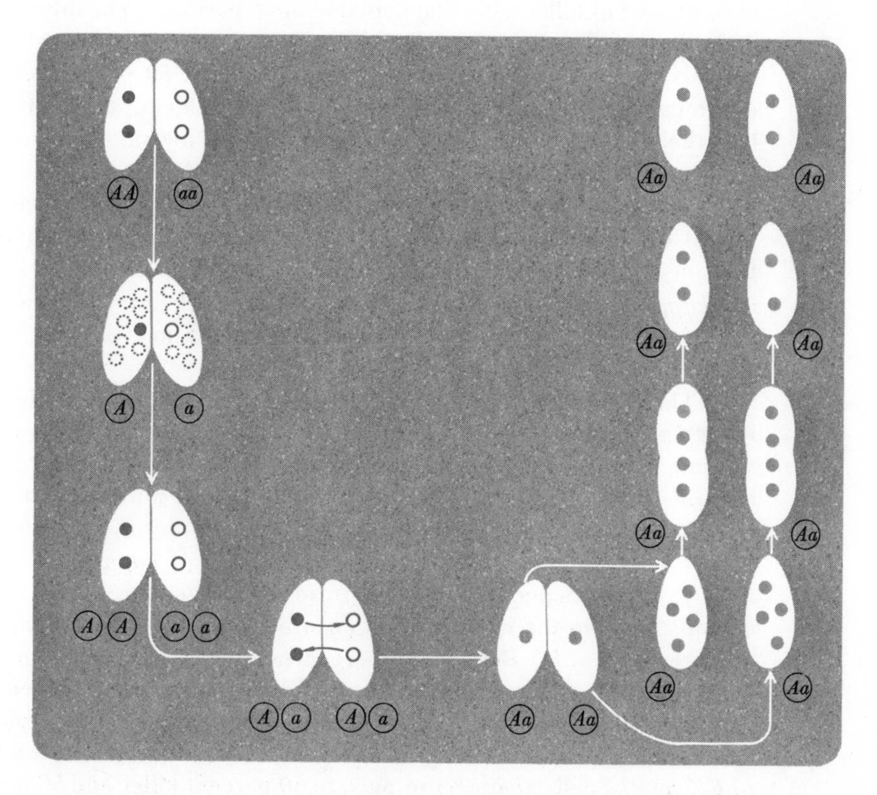

Fig. 7-2 Sexual cycle in Paramecium aurelia.

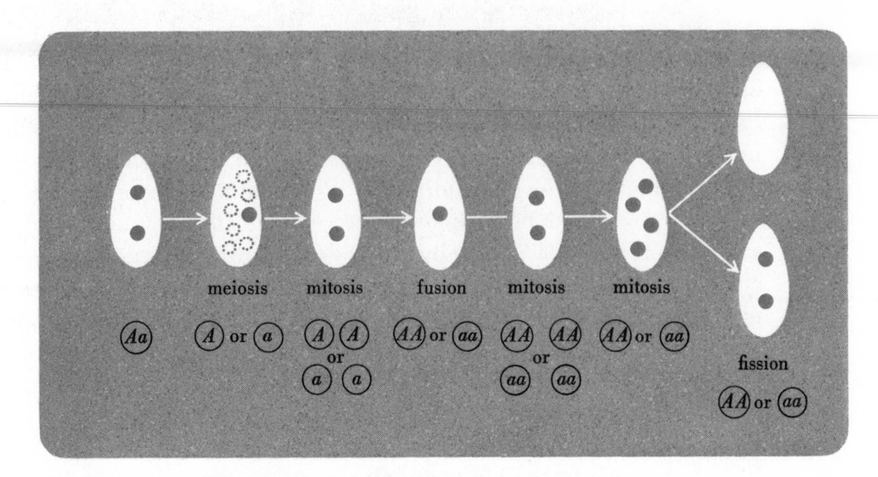

Fig. 7-3 *Autogamy in* Paramecium aurelia.

duction of two kinds of clones. One clone, derived from the killer cell, is composed of killer cells that contain *kappa* particles. The other clone, derived from the sensitive cell, is all sensitive and lacks *kappa*. When animals from the killer clone undergo autogamy, 50 percent of their exautogamous progeny possess *kappa* and are killers, and 50 percent lose their *kappa* and become sensitives; among the sensitive animals, autogamy leads only to sensitive progeny (Fig. 7-4).

When killers and sensitives are crossed under conditions that permit cytoplasmic exchange to occur, a different result is obtained. Both of the clones derived from the two conjugating cells are killers, and *kappa* particles are found in the cells of both clones. Autogamy also yields a different result, since it gives 50 percent killers and 50 percent sensitives from both clones (Fig. 7-5).

The results of these crosses between killer and sensitive animals, $KK \times kk$, show that the progeny are all *Kk* in genotype. However, as we saw in the first cross, only the clone derived from the killer parent contains *kappa* particles. When these cells undergo autogamy they give rise to *KK* and *kk* cells; the former have *kappa* and the latter do not (Fig. 7-4). This indicates that the maintenance of *kappa* particles in the cytoplasm depends upon the presence of the gene *K*, as further demonstrated in the second cross, in which cytoplasmic exchange has occurred (Fig. 7-5). Both clones are genotypically *Kk* as before, but cytoplasmic exchange has transferred *kappa* particles to the sensitive cell. *Kappa* particles are maintained in the progeny derived from this cell because these cells have the *K* gene. Autogamy once again leads to *KK* and *kk* cells and, accordingly, to 50 percent killer and 50 percent sensitive cells. It can be shown, however, that after autogamy of *Kk*

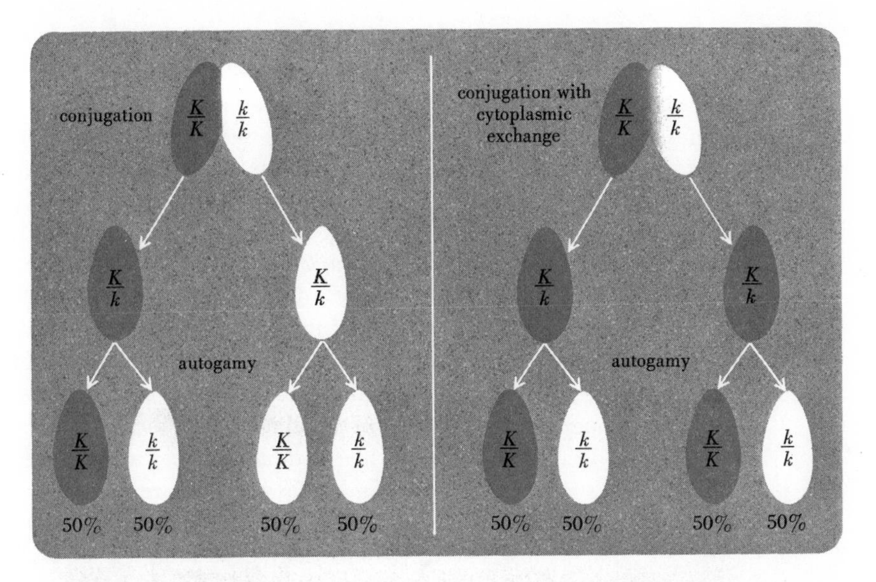

Fig. 7-4 Paramecium aurelia: *Conjugation between killer and sensitive animals without cytoplasmic exchange, followed by autogamy.*

Fig. 7-5 Paramecium aurelia: *Conjugation between killer and sensitive animals with cytoplasmic exchange, followed by autogamy.*

killers, the exautogamous *kk* animal does contain *kappa* particles, but after a few generations of fission the particles are lost. Thus, the particles cannot reproduce unless a *K* gene is present.

Killer strains have been isolated that possess particles morphologically different from *kappa* and that produce killing substances with different properties. It is therefore inferred that *kappa* particles are capable of undergoing mutation and that these mutations are replicated and result in different phenotypes.

Though the *kappa* particles are extranuclear, their reproduction depends upon the genotype of their host. It is difficult to decide whether *kappa* particles should be considered as normal particles within the cells of *P. aurelia* or whether they are symbiotic. Regardless, they are self-duplicating, DNA-containing particles that are transmitted from generation to generation.

The examples of extranuclear inheritance in *Drosophila* and *Paramecium* appear to be somewhat exceptional in that they deal with unusual cytoplasmic constituents. *Sigma* particles have some of the properties of viruses, whereas *kappa* particles may be similar to certain bacteria. It is difficult in these two cases to state that they are either normal or abnormal cytoplasmic constituents.

ORGANELLE HEREDITY If we turn from what may seem to be two unusual examples of extranuclear inheritance, we will find others that involve the inheritance of normal cytoplasmic bodies, such as the chloroplasts and mitochondria or their constituents. The chlorophyll of green plants is localized in specialized structures called chloroplasts. The chloroplasts of most green plants arise from the development of plastid primordia, and in higher plants these primordia are transmitted by the female parent. They are found within the embryo sac, and are thus constituents of the cytoplasm in which the embryo will develop. The number of such primordia transmitted is quite small, and thus the millions of chloroplasts in a mature plant must arise after zygote formation. Do these chloroplasts arise *de novo* or through self-duplication of the maternally inherited primordia?

Several different species of higher plants have variegated leaves; the leaves are a mixture of green and yellow leaf tissue. In certain cases one plant can exhibit branches with green leaves, with variegated leaves, or with yellow leaves. Microscopic examination of the green leaves, and green areas of variegated leaves, reveals that the cells contain normal chloroplasts and chlorophyll pigment, whereas the yellow leaves and yellow areas of variegated plants lack normal chloroplasts and pigment. The progeny from variegated plants reveal how the chloroplasts are inherited (Table 7-1).

It will be seen from Table 7-1 that the kind of chloroplasts that the progeny have depends upon the kind possessed by the female parent. Females from green portions of the plant transmit only plastid primordia capable of developing into normal, green chloroplasts, whereas females from yellow parts of the plant transmit only abnormal plastid primordia that will not develop pigment. On the other hand, flowers from the variegated regions of the plant have the two kinds of

Table 7-1 **Chloroplast Inheritance in Variegated Four-O'Clocks**

Branch of Origin of the Male Parent	*Branch of Origin of the Female Parent*	*Progeny*
green	green	green
	pale	pale
	variegated	green, pale, variegated
pale	green	green
	pale	pale
	variegated	green, pale, variegated
variegated	green	green
	pale	pale
	variegated	green, pale, variegated

primordia, and an embryo sac may receive only normal primordia and produce green offspring. The embryo sac, however, may receive only abnormal primordia and, as a consequence, the plants will be yellow. If the primordia in the embryo sac have both types, a variegated plant will result. This inheritance of chloroplast type appears to be independent of the nuclear genotype of either the male or female parent.

The two sorts of chloroplasts described here are relatively stable types that are reproduced and transmitted through many plastid generations. It is considered likely that the abnormal chloroplasts arise from normal chloroplasts by mutation. M. M. Rhoades has investigated stable plastid mutations in maize, and has shown that such mutations occur in the presence of a specific nuclear gene, *iojap (ij).* When plants are homozygous for this recessive gene, the mutation rate from normal to abnormal plastids is greatly increased. Once the plastid mutation has occurred, however, it is stable, and its transmission is not modified by the presence or absence of the nuclear *ij* gene.

Recent investigations of chloroplast inheritance have led to the conclusion that the chloroplasts themselves possess certain genetic properties. To understand the nature of some of these properties it has been necessary to turn to investigations with microorganisms because of their rapid generation time and because various experimental techniques can be used that are difficult with higher plants. To date, one of the most useful microorganisms for these investigations has been the unicellular green alga *Euglena.* When cultured in the light, cells of this alga are green and have approximately 12 chloroplasts. When cultured in the dark, the cells are albino and instead of chloroplasts they have the precursors or primordia from which chloroplasts will develop. If dark-grown albino cells are placed in the light they will develop chloroplasts and become green. All succeeding generations of these cells, when cultured in the light, will have chloroplasts and be green. The dark-grown albino cells are thus *reversibly bleached.*

In contrast to normal cells that are reversibly bleached, it is possible to obtain strains of *irreversibly bleached* cells by exposing green cells to irradiation with x-rays or ultraviolet (UV) light, to streptomycin, or to temperatures of 32 to 34°C.

When green cells are exposed to a dose of UV light that does not affect their viability, 100 percent of the irradiated cells give rise to albino colonies when grown in the light. If albino cells, obtained after green cells have been grown in the dark, are irradiated with UV and then cultured in the light again, 100 percent of the colonies are albino. In either case, subcultures of the albino cells give rise only to albino colonies, and for many generations no reversion to green cells occurs. Thus, the UV irradiation has brought about an inheritable change, which can be induced either in light-grown cells that have chloroplasts or in dark-grown cells that have only the precursors of chloroplasts.

The effect of UV irradiation of *Euglena* has been localized to the cytoplasm. When single green cells are exposed to a microbeam of UV that irradiates the entire cell, the progeny are albino. They are also albino when UV irradiation is given to cells in which the nucleus is shielded. However, if the microbeam is focused on the nucleus of a single cell all of the cell's progeny are green. Therefore, the effect of the UV is upon some component in the cytoplasm rather than on the nucleus. Furthermore, the wavelengths of 2600 and 2800 Å have been found to have the maximum effect. Since nucleoprotein absorbs the maximum amount of UV at these two wavelengths it can be tentatively concluded that the cytoplasmic component is, or contains, a nucleic acid.

During the past several years the technique of density-gradient centrifugation (see Chapter 1) has established that chloroplasts contain DNA, which is distinct from nuclear DNA. When an extract of green or reversibly bleached *Euglena* cells is centrifuged, three bands of DNA are readily detected by their differences in density. One band, the major band, is the nuclear DNA, whereas the two minor or satellite bands correspond to some other DNA-containing components of the cell. Now, if a cell extract containing only chloroplasts is prepared and then centrifuged in the density gradient, a single band of DNA is seen with a density that corresponds to one of the satellite bands. This observation suggests that the chloroplasts contain DNA. Further substantiation from chemical and cytological methods demonstrates the presence of DNA in the chloroplasts. The second satellite band represents mitochondrial DNA.

You will recall that strains of irreversibly bleached cells can be produced by several different means. Unlike cells of reversibly bleached strains, none of these strains possess the satellite band of DNA associated with the chloroplasts. These strains also lack chloroplasts and the primordia from which chloroplasts can develop.

Although DNA has been shown to be a component of the chloroplasts of many lower and higher plants, its function is not yet fully understood. Experiments with *Euglena* done by J. A. Schiff and his colleagues have provided some information regarding its function. In order to interpret the results of the experiments, however, it is necessary to digress somewhat to discuss a few details regarding the UV irradiation of *Euglena*. It was stated earlier that when green cells are irradiated with UV, 100 percent of the colonies that arise are albino. However, in order to obtain this result it is necessary to incubate the irradiated cells in the dark for several days before placing them in the light. If the treated cells are placed in the light immediately after UV irradiation, 100 percent of the colonies that arise are *green*. This result is obtained because of a chemical process known as *photoreactivation*, which "repairs" the damage caused by UV irradiation.

Another phenomenon is the effect of light on the formation of green cells from dark-grown albino cells. The principal wavelengths of light responsible for photoreactivation are between 3300 and 4600 Å. However, chloroplast formation can occur efficiently in red light at wavelengths from 6000 to 7000 Å, whereas photoreactivation does not. Thus, it is possible to irradiate cells with UV and to incubate them immediately in light that permits chloroplast development without permitting photoreactivation. Then, after a few days, the colonies can be placed in white light. The use of red light to induce chloroplast formation makes it possible to investigate the effects of UV on chloroplast duplication versus chloroplast development.

In the Schiff experiment, green cells were irradiated with UV and placed immediately in the dark, in white light (photoreactivating), or in red light (nonphotoreactivating) for 72 hr on a medium that permitted the cells to survive but not to divide. After 72 hr, samples of cells from each of the three conditions were divided into two groups. One group was placed in white light for seven days and the other in the dark for five days followed by two days in the light. Cells that did not receive UV irradiation were treated in the same way and served as a control.

Chloroplast formation among UV-treated *cells* was found to be normal when they were exposed to either red or to white light for 72 hr. Thus, the UV irradiation did not affect chloroplast development from the chloroplast precursors in the cells, even when they were kept under nonphotoreactivating conditions (red light). However, when the irradiated *cells* that developed chloroplasts under nonphotoreactivating conditions (red light) were permitted to divide many times to form colonies, 100 percent of the *colonies* were found to be albino. Thus, the albino colonies developed from cells that were themselves capable of forming chloroplasts. Therefore, it was evident that the UV irradiation did not destroy the cell's ability to develop chloroplasts; rather, it destroyed their ability to *replicate* the chloroplasts.

The conclusion drawn from the observations presented here is that some cytoplasmic component, most likely chloroplast DNA, controls chloroplast replication but does not necessarily affect chloroplast formation from the chloroplast precursors already present in the cell. This cytoplasmic component can be inactivated with UV so that its duplication does not occur. Treated cells form chloroplasts but do not replicate chloroplasts and do not transmit them from one cell generation to the next.

DNA has been found to exist not only in chloroplasts, but also in mitochondria. For example, *Neurospora* mitochondria contain DNA that is distinct from the nuclear DNA. There are several different mutant strains of *Neurospora* showing maternal inheritance, that possess abnormalities associated with either mitochondrial function or structure. However, the role that mitochondrial DNA may play in determining either the inheritance or function of mitochondria remains to be demonstrated.

Organelle DNA must play some genetic function. One function may be to determine the amino-acid sequence of the structural proteins of which the organelle is constructed, and another may be to determine the amino-acid sequence of any one of several enzymes that are essential for the function of the organelle. These and other functions for organelle DNA are being sought. All that can be stated at present is that chloroplast DNA has at least one function: that of playing a role in chloroplast replication.

INHERITANCE OF PREFORMED STRUCTURE

Infectious heredity and organelle heredity can be attributed to DNA that is found outside of the nucleus either in infectious particles or in organelles. There are, however, hereditary characteristics whose origin is not easily ascribed to either nuclear or extranuclear DNA, and these characteristics are certain of the structures of the cortex of ciliated protozoa such as *Paramecium.* The mouth and the contractile vacuole are two prominent features of the cortex of *Paramecium*, and Sonneborn and his co-workers have shown that these existing, or *preformed*, structures can be transmitted from one cell generation to the next independently of the transmission of nuclear genes and cytoplasmic genetic factors.

During sexual reproduction in *Paramecium* (Fig. 7-2), two cells conjugate and then separate after they have exchanged nuclei. However, on occasion the conjugants fuse and a double, or doublet, animal is formed having two sets of cortical structures. When doublets reproduce asexually by fission they give rise to doublets. This shows that there is a genetic basis for the double nature of the cortical structures; but significantly, when doublets are mated with normal, or single, animals the progeny of the doublet exconjugant are doublets and the progeny of the singlet exconjugant are singlets (Fig. 7-6). Also, when the progeny go through autogamy they maintain their doublet or their singlet properties. However, nuclear gene markers introduced into the crosses between doublets and singlets are transmitted and segregate according to the pattern expected for nuclear genes.

The mode of inheritance of the duplicated structures, therefore, appears to be independent of the mode of inheritance of nuclear genes. It is also independent of cytoplasmic hereditary factors. *Kappa* particles can be used to mark the cytoplasm of either doublet or singlet

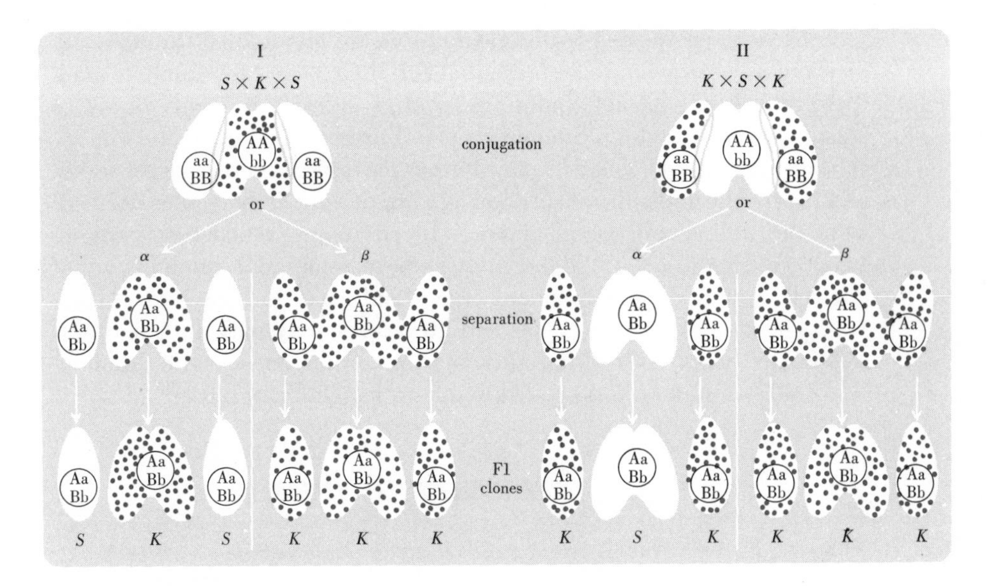

Fig. 7-6 *Cross I: Killer doublet × sensitive singlets. Cross II: Sensitive doublet × killer singlets, α, conjugation without transfer of cytoplasm; no cytoplasmic bridges between mates. β, conjugation with transfer of cytoplasm via cytoplasmic bridges connecting mates.* A *and* a, B *and* b *are two pairs of allelic genes.* S, *sensitive;* K, *killer; dots represent kappa in cytoplasm. (After Sonneborn, in* The Nature of Biological Diversity, *J. M. Allen, ed. New York: McGraw-Hill, 1963.)*

animals, and matings can be made between doublets that have *kappa* and are killers and singlet animals that do not have *kappa* and are sensitives. The reciprocal cross can also be made. If conditions for cytoplasmic exchange are permitted, its occurrence will be marked by the transfer of *kappa* particles (Fig. 7-6).

When the matings are made it is found that the transmission of kappa as well as that of the killer and sensitive phenotypes follows the expected pattern. However, the doublet and singlet exconjugants reproduce their kind among their progeny. These in turn reproduce true to type following autogamy. These observations reveal that the inheritance of the cortical structures is not affected by exchange of cytoplasm and is apparently independent of cytoplasmic genetic factors.

An additional observation supporting the idea that the cortical structures are autonomous concerns the natural grafting of a piece of cortex from one conjugant to the other. This grafting happens only rarely, but when it does an animal is recovered that has duplicated cortical structures. When this animal reproduces, the new structures reproduce autonomously.

Clearly, preformed cortical structures are maintained through cell division; they appear to be essential for their own reproduction, and their inheritance is not under the control of nuclear genes or cytoplasmic genetic factors. Sonneborn (see Further Reading at the end of this chapter) has suggested that different parts of the cortex might serve as sites for the "specific absorption and orientation of molecules derived from the milieu and genic action." In addition, pre-existing cortical structures could act by "determining where some gene products go in the cell, how these combine and orient, and what they do."

The inheritance of preformed structures may play an important role in the course of cell heredity and in morphogenesis. The problem is to seek an understanding of this role in molecular terms.

FURTHER READING

Ephrussi, B., 1953. *Nucleo-cytoplasmic relations in microorganisms*. New York: Oxford University Press.

Jinks, J. L., 1964. *Extrachromosomal inheritance*. Englewood Cliffs, N.J.: Prentice-Hall, Inc.

Sonneborn, T. M., 1963. "Does preformed structure play an essential role in cell heredity?" in *The nature of biological diversity*. Allen, J. M. (ed.). New York: McGraw-Hill Book Company, Inc., pp. 165–221.

Wilkie, D., 1964. *The cytoplasm in heredity*. London: Methuen & Co., Ltd.

chapter 8

The Transmission of the Genetic Material in Bacteria and Bacterial Viruses

We have seen the relationships between the meiotic transmission of the genotype and segregation, independent assortment, and linkage and crossing over. These phenomena are all reflections of the process of sexual reproduction. They are characteristic of the transmission of the genetic material in all organisms possessing a nucleus and chromosomes. However, there are organisms, such as the blue-green algae and the bacteria, that are nonnucleate. There are also the bacterial, plant, and animal viruses whose mode of life and reproduction seem to place them in a class by themselves. There is no evidence to show that the bacteria or viruses possess a classical type of meiosis; yet there is no evidence to show that they do not. Certain distinctive characteristics of the transmission of their genotypes, however, appear to set them apart from nucleate organisms. In view of the absence of a clearly defined cytological process akin to meiosis, we will consider the transmission of their genotype as being nonmeiotic. In this chapter we will inquire into the nature of their genetics and some of the ways in which their genetic material is transmitted from one cell generation to another.

Three different modes of transmission of the genetic material have been discovered in bacteria. They are conjugation between cells of opposite mating type, transformation (mentioned briefly in Chapter 1), and transduction, a form of genetic transmission in bacteria that is mediated by certain bacteriophages. All three can yield progeny with recombinant genotypes, all involve *donor* and *recipient* bacteria, and normally there is only a *partial* transfer of genetic material from a donor to a recipient. The three modes also require an even number of recombinational events in order to obtain recombinant progeny.

CONJUGATION AND GENE RECOMBINATION IN ESCHERICHIA COLI

In 1946 J. Lederberg and E. L. Tatum investigated whether there was any form of sexual reproduction in strain *K*-12 of the bacterium *E. coli*. They utilized two mutant strains that Tatum had obtained by the use of x-rays and ultraviolet light (Chapter 9). Wild-type *E. coli* grows on a minimal medium containing inorganic salts and a source of carbon such as glucose, and from these materials wild-type *E. coli* can synthesize all the substances necessary for growth and reproduction. The mutant strains of *E. coli*, strain *K*-12, used by Lederberg and Tatum could not grow on minimal medium, since each was unable to synthesize all of its necessary vitamins and amino acids. They grew, however, on suitably supplemented media. One strain (*Y*-10) required the amino acids threonine and leucine, and the vitamin thiamine for growth, and the other strain (*Y*-24) required the amino acids phenylalanine and cystine and the vitamin biotin. Each strain, therefore, carried three different biochemical mutations. It was observed that either of the strains could give rise to colonies that were wild type with respect to *one* of these biochemical mutants at the low frequencies of one per million (1×10^{-6}) to one per hundred million (1×10^{-8}) bacteria. These new types arose by mutation just as in the case of the rough pneumococci that arose in cultures of smooth pneumococci (Chapter 1). The occurrence of a change from mutant to its wild-type alternative, such as seen in the *E. coli* strains, is called *reverse mutation.*

Since the frequency of reversion was of the order of 1×10^{-6} to 1×10^{-8} for any *one* of the mutants present in the two strains, the frequency of the simultaneous reversion of two or three of them is so many orders of magnitude smaller as to become insignificant. For example, if in strain *Y*-10 the threonine and leucine mutants each have a reversion rate of one in 10^6 bacteria, their simultaneous reversion would be the product of the two events or one in 10^{12} bacteria.

When Lederberg and Tatum grew *mixed* cultures of strains *Y*-10 and *Y*-24, they found colonies that were able to grow on minimal medium. These colonies must, therefore, have been wild type for all *six* of the mutants present. In addition, they found colonies that had only a single requirement for one of the amino acids or vitamins, as well as various combinations of the six mutant types. In other words, they observed an array of *new* gene combinations. The frequency of occurrence of the new gene combinations or recombinants was of the order of one per million cells. The fact that some of the recombinants involved all or several of the mutants rules out their origin by simultaneous reverse mutations. Lederberg and Tatum concluded, therefore, that the new types arising in the mixed cultures were the result of a process that permitted a recombination of genetic material.

From their findings they inferred the existence of some form of sexual process in *E. coli*. In addition, they hypothesized that the two parental cells fused, that a diploid zygote was produced, and that a process akin to meiosis yielded the recombinant types. However, subsequent experiments by J. Lederberg, F. Jacob and E. L. Wollman, W. Hayes, E. A. Adelberg, and others led to a revision of this hypothesis and provided a wealth of information regarding the genetics of *E. coli*. Today the genetic systems of *E. coli* and certain related bacteria are well understood and have become as "classical" as those of *Drosophila*, maize, and mice. Only a few of the elegant experiments that led to the elucidation of the genetics of *E. coli* can be considered here, and the reading list at the end of this chapter will aid you in a further understanding of the genetic systems of bacteria.

Early in the investigation of the genetics of *E. coli* it became apparent that the transmission of the genetic material differed from that encountered among organisms possessing a classical meiosis. For example, the two parental types did not make equal genetic contributions to their progeny, and certain combinations of parental genes appeared in the progeny whereas others were always missing. In addition, all of the genes found among the progeny were linked. These peculiarities became understood only after Hayes demonstrated that sexual differentiation occurs in *E. coli*.

Certain crosses between different strains of *E. coli* are fertile (that is, they yield recombinants), whereas others are sterile and no recombinants are found. This result suggests some form of mating type differentiation in which cells of opposite mating type are required in order to obtain recombinants. Furthermore, it was shown that one parent acts as a *donor* of genetic material and that the other parent is a *recipient*. Two different strains were used by Hayes to demonstrate this point. Strain A required the amino acid methionine (*met*$^-$) and the vitamin biotin (*bi*$^-$) for growth, and strain B required the amino acids threonine (*thr*$^-$) and leucine (*leu*$^-$) and the vitamin thiamin (*thi*$^-$). When crossed, these two strains yielded recombinants that grew on minimal medium. Next, an additional pair of genetic markers, streptomycin sensitivity (*str-s*) and resistance (*str-r*), was introduced into the cross. When strain A carried the gene *str-s* and strain B carried its allele, *str-r*, recombinants were obtained at the same frequency when the mixture of the two strains was plated either to minimal medium containing streptomycin or to medium without streptomycin. However, when strain B was *str-s* and strain A *str-r*, no recombinants were found on the streptomycin medium. In another experiment, cells of strain A or B were treated with streptomycin before being crossed. They were exposed to the antibiotic for up to 18 hours in order for it to exert its effect. Then they were washed to remove the antibiotic and mixed with the appropriate untreated cells. Next they were plated to minimal medium, and it was found that no recombinants were produced when the *str-s* strain B was exposed to streptomycin. Hayes concluded that the survival of cells of strain B was required for fertility and that zygote formation and recombination occurred within strain B cells. Thus, the survival of cells of strain A was not required for zygote formation but only for the transfer of the genetic material to the B cells. Accordingly, the B cells were the recipients and the A cells the donors of the genetic material.

Other evidence demonstrates that there are donor and recipient cells. It has been shown that during mating, DNA labeled with an isotope is transferred in only one direction, and when conjugating bacteria are separated by micromanipulation, the recombinant types of progeny are found to arise from only *one* of the exconjugants. Mixtures of cells of recipient strains do not yield recombinants, and are always infertile or sterile. These recipient cells are given the designation F^-, where the minus sign indicates that these cells lack a fertility or sex factor, called the *F* factor. When donor strains are mixed with either recipients or *other donors*, fertile crosses occur; that is, recombinants are formed, even though the frequency of recombinants is relatively low in crosses between donors. Donor strains possess the *F* factor and are designated F^+.

It will be recalled that recombinant types from a cross between donor and recipient bacteria arise at a frequency of about 1×10^{-6}. It is therefore evident that the F^+ strain does not transfer its genetic material very efficiently. However, the *F* factor itself is transferred to the recipient bacteria with high efficiency, for within an hour after mixing the F^+ and F^- cells, 95 percent of the F^- cells have become converted into F^+ cells. In other words, the fertility or sex factor can be transmitted to recipient cells independently of the transfer of the chromosome. The F factor has been identified as a component of the cytoplasm of the donor cells, and it contains an amount of DNA equivalent to 2.5×10^5 nucleotide pairs. When a large number of F^- cells is mixed with a small number of F^+ cells, the F^- cells are converted to F^+ at a rate that exceeds the rate of cell division. Therefore, the F factor can reproduce autonomously.

The F factor possesses another and somewhat remarkable property. In addition to its ability to be transmitted by infection, the factor can become integrated into the bacterial chromosome at a specific site. In this case its reproduction is no longer autonomous but becomes associated with the duplication of the bacterial chromosome, and it loses its capacity to infect F^- cells. The F factor and certain other genetic elements that have been discovered can exist either autonomously in the cytoplasm or in association with the bacterial chromosome. They are called *episomes*. When cells in which the F factor becomes associated with the bacterial chromosome are mixed with F^- cells, recombinants are now recovered at frequencies of 1×10^{-3} to 1×10^{-1}. Recombination is thus much more frequent than the 1×10^{-6} found for $F^+ \times F^-$ crosses. In crosses with a high frequency of recombination, the donor strains are termed *Hfr* (for high frequency of recombination).

The initial step in genetic transmission in *E. coli* involves effective contact between cells of opposite mating type. It is known that effective contact is established quite rapidly; in most cases, the mating process is completed in 30 min or less.

The second step in genetic transmission is the transfer of genetic material *from the Hfr donor to the* F^- *recipient in an oriented fashion.* In order to understand this very fascinating and intricate step in the process let us see what occurs when a cross is performed between genotypically different *Hfr* and F^- strains. The *Hfr* strain in this cross has the following genotype: thr^+ leu^+ *azi-s* T_1-*s* lac^+ gal^+ *str-s* (*Hfr*). It is capable of synthesizing the amino acids threonine (thr^+) and leucine (leu^+); it is sensitive to the metabolic inhibitor sodium azide (*azi-s*) and sensitive to the bacteriophage known as T_1 (T_1-*s*). In addition to its ability to ferment glucose, it can also ferment lactose (lac^+) and galactose (gal^+), and finally, it is sensitive to the antibiotic streptomycin (*str-s*). The F^- strain has the complementary genotype, that is, thr^- leu^- *azi-r* T_1-*r* lac^- gal^- *str-r* (F^-). A cross is made by mixing the *Hfr* and F^- cells and letting them remain together for 25 min. After mixing, the cells are plated to minimal medium containing streptomycin. Disregarding *all but* the *thr*, *leu*, and *str* loci for the moment, it is evident that the only cells that will be capable of forming colonies on this medium will have the recombinant genotype *thr*$^+$ *leu*$^+$ *str-r*. The thr$^+$ and leu$^+$ are from the *Hfr* parent, and *str-r* is from the F^- parent. This initial step of selection of recombinant genotypes is made because the frequency of recombinants is much less than either parental type. The *Hfr* parent will be killed because the medium contains streptomycin, and the F^- parent will be unable to grow since threonine and leucine are absent.

When the *thr*$^+$ *leu*$^+$ *str-r* colonies are replica plated to a minimal medium containing either sodium azide or bacteriophage T_1, the genotypes for the *azi* and T_1 loci can be determined. Furthermore, replica plating to special media will indicate their ability to ferment either lactose or galactose. Results typical of such crosses have been reported by F. Jacob and E. L. Wollman. Among the *thr*$^+$ *leu*$^+$ *str-r* recombinants they found that 90 percent carried the *azi-s* marker of the *Hfr* donor. The other markers contributed by the *Hfr* were T_1-*s*, 70 percent; *lac*$^+$, 40 percent; and *gal*$^+$, 25 percent. In other words, when a *thr*$^+$ *leu*$^+$ *str-r* recombinant was formed, 90 percent of the time it was a *thr*$^+$ *leu*$^+$ *azi-s* *str-r* recombinant; 70 percent of the time it was a *thr*$^+$ *leu*$^+$ *azis-s* T_1-*s* *str-r* recombinant; and so forth. See line A of Table 8-1.

The transfer of genetic material from *Hfr* to F^- appears to be directional or oriented. The oriented nature of the genetic transmission from *Hfr* to F^- has been shown by experiments in which the mating process has been interrupted at different times, followed by genetic analysis correlating the genetic contribution made by the *Hfr* parent with the time of interruption of conjugation. Jacob and Wollman interrupted mating by separating the conjugants by agitating them in a Waring Blendor.

Results of an experiment in which mating was interrupted are shown in line B of Table 8-1. The percentage of each of the recombinants between *Hfr* and F^- (line A) is compared with the time of its appearance after the interruption of mating. Several facts can be ascertained from these results. First, the experiment shows that the *Hfr* parent transfers its chromosome to the F^- parent and that it donates the chromosome segment in a sequence from *thr* to *gal*. Second, the disruption of mating at different times shows that the transfer of genetic material is oriented, beginning with the *thr* locus, then the *leu* locus, then *azi*, and so forth. The *thr* locus, therefore, is close to the part of the chromosome that is transferred first (point of origin). It has been found that the *str-s* of this *Hfr* strain is not transferred. Third, the sequence of the transfer of the loci is in agreement with data

Table 8-1 **Genetic Map of the Segment Injected with High Frequency by *Hfr E. coli***

	0	*thr*$^+$	*leu*$^+$	*azi-s*	T_1-*s*	*lac*$^+$	*gal*$^+$
Percentage of Recombinants	A	–	–	90	70	40	25
Time (minutes)	B	8	8½	9	11	18	25

obtained in a genetic analysis; there is a close relationship between the time of recovery of a given gene from *Hfr* after interruption, and the frequency with which this gene is recovered when recombination is studied in crosses where mating is not interrupted (Table 8-1). For example, a gene close to the origin, such as *azi*, is also one which is found in high frequency among the recombinants. The Waring Blendor experiment shows that *azi-s* is transferred as early as 9 min. It is also found in 90 percent of the recombinants in experiments in which mating is allowed to proceed in a normal fashion.

A number of different *Hfr* strains have been obtained from F^+ bacteria, and each strain transfers markers at a high frequency. Each is distinguished by having a different origin, and thus a different genetic map based upon the orientation of the transfer of the genetic material to the F^- cell. The order of transfer characteristic of several different *Hfr* strains is given in Fig. 8-1. The relationship between individual genes, however, remains the same regardless of the site of the origin and the direction in which the segment is transferred. Furthermore, all of the genes are linked in a single linkage group.

The *Hfr* strains, each showing a characteristic pattern of chromosome transfer, have been shown to arise from F^+ cultures, and different *Hfr* strains can be recovered from the same culture of F^+ bacteria. An analysis of a variety of crosses involving different *Hfr* strains reveals that, with rare exceptions, only a part of the chromosome is transferred to the recipient cell. The reason is that the donor chromosome usually breaks before all of it is transferred. Thus, the further a given marker is from the origin, the smaller is the probability of its transfer. Although it is always possible to determine the point of origin for a given *Hfr* strain, it is not always possible to determine the distal end of the chromosome. However, if the proper precautions are taken during the course of mating the entire chromosome may be transferred; in this case the *Hfr* property itself is transmitted to the progeny, and they can now act as donors.

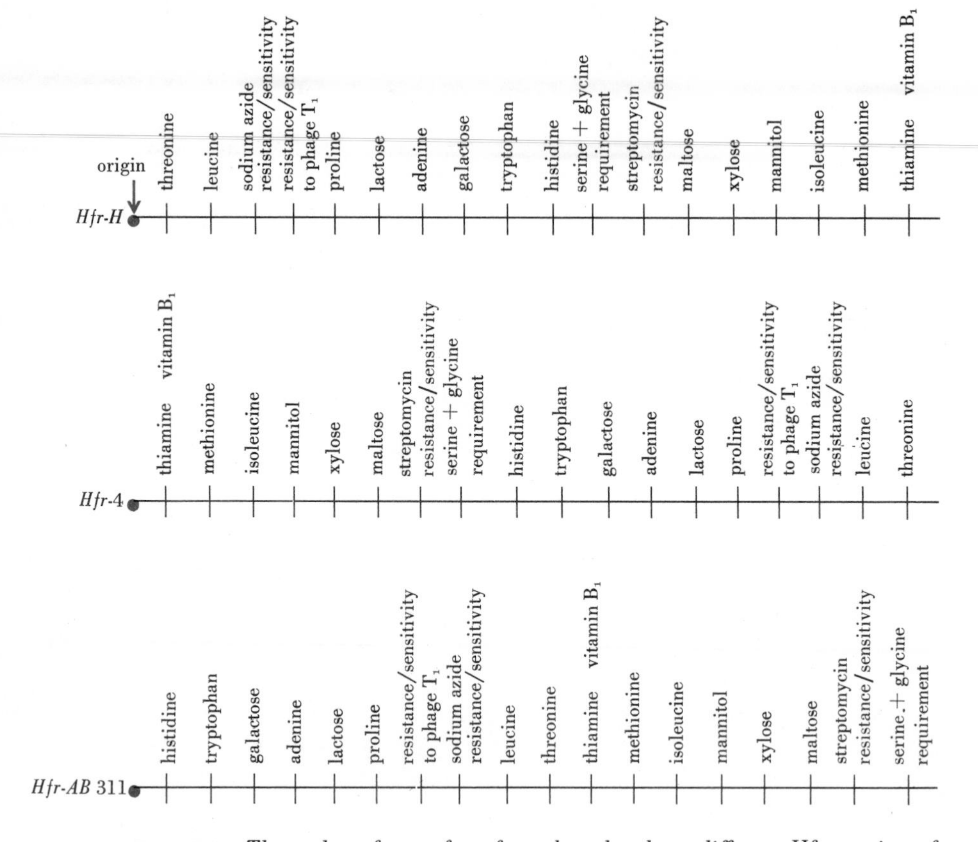

Fig. 8-1 *The order of transfer of markers by three different* Hfr *strains of* E. coli. *The origin is indicated on the left.*

The relationship between the sex factor *F* in F^+ and *Hfr* cells can be visualized in the following manner. *F* can exist in two alternative states: it can become integrated into the chromosome (*Hfr* cell) or it can remain free (F^+ cell). The point at which the transfer of the donor chromosome begins (the origin) is determined by the position that *F*

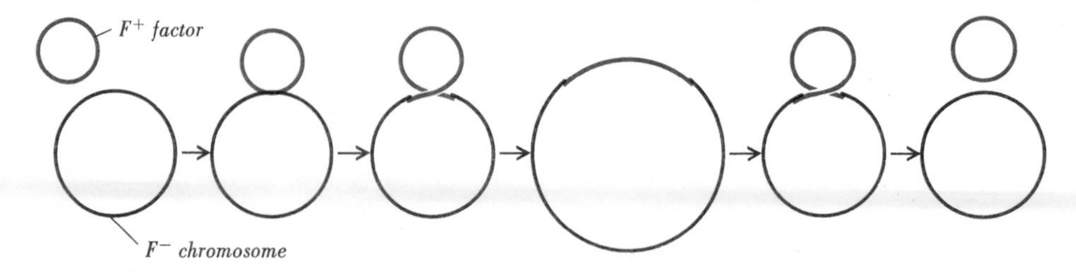

Fig. 8-2 *Diagrammatic representation of how the* F$^+$ *factor becomes integrated into (or separated from) the* F$^-$ *chromosome by a single crossover event.*

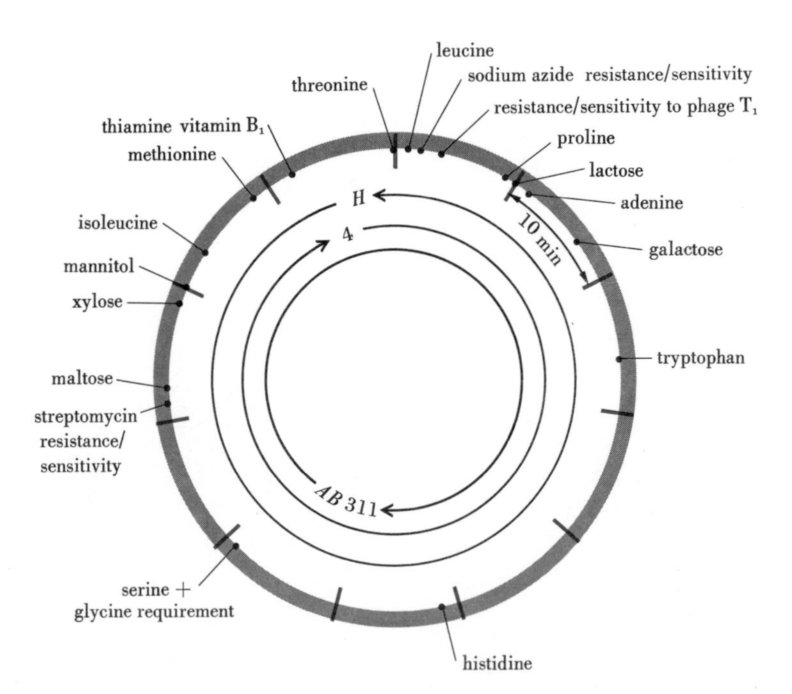

Fig. 8-3 *The circular genetic map of* E. coli. *The chromosome is divided into eleven intervals, each corresponding to 10 min of transfer time. The order of transfer is shown for the three different* Hfr *strains depicted in Fig. 8-1. An arrowhead indicates the origin. The circular map does not include all of the known markers.*

takes in the chromosome. Jacob and Wollman suggested that the chromosome of an F^+ cell is a circle; subsequently it was suggested that the F factor is also circular. A. Campbell proposed that the F factor is integrated into the chromosome by a single reciprocal crossover (Fig. 8-2). Furthermore, Campbell proposed that a single reciprocal crossover can release the F factor from a chromosome. When the F factor becomes integrated, the chromosome breaks at the point of integration.

When data from the transfer of chromosomes from different *Hfr* strains are combined, a circular chromosome map, as shown in Fig. 8-3, is obtained. This genetically circular chromosome, based solely upon the analysis of crosses, was later shown to be physically a circle in both F^+ and F^- strains (Fig. 8-4).

The next step in the transmission of the genetic material in *E. coli* is the series of events that result in the *integration* of the transferred *Hfr* markers into the recipient's genome. The chromosome from the donor pairs with that of the recipient cell, and if a minimum of

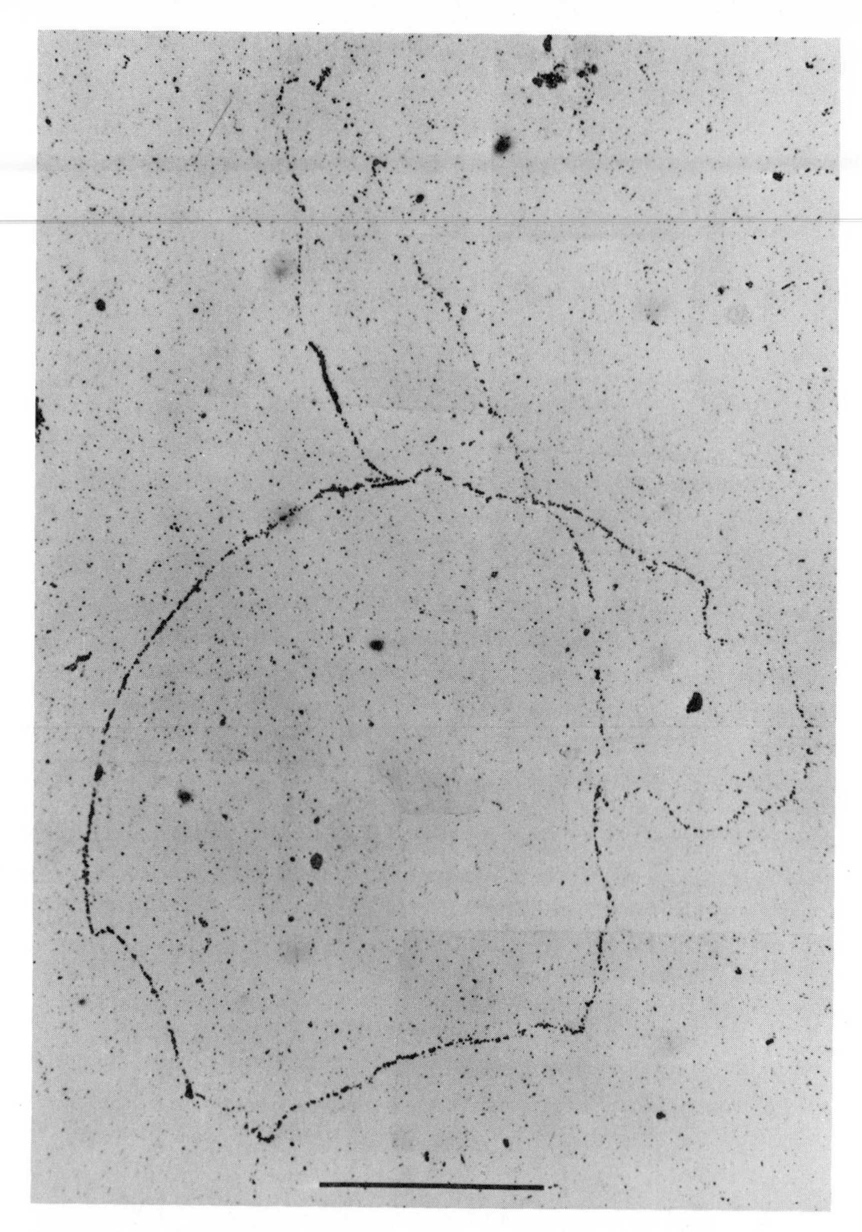

Fig. 8-4 *Autoradiograph of the replicating, circular chromosome of* E. coli, *extracted after labeling for about two generations with H^3-thymidine. The scale shows 100μ. (Dr. John Cairns, from Cold Spring Harbor Symposium of Quantitative Biology, 1963.)*

two recombinational events occurs, the recipient will carry a recombinant chromosome. A selection procedure is then used by the investigator to recover the desired types of recombinants.

Finally, let us consider the question of making a genetic map for a portion of an *E. coli* chromosome transferred by a given *Hfr* strain. It should be apparent by now that there are two sorts of units that can be used to construct a genetic map of *E. coli:* time units and recombination units. A map constructed on the basis of time units can be obtained by interrupting the mating process at specific times and then plating the cells to the appropriate selective media. This map, with distances in time units, also gives the relative positions of the genes in the chromosome (see Fig. 8-3). However, mapping by time units has the disadvantage that it is not possible to obtain accurate information with regard to the distances between closely linked loci such as *thr* and *leu*; in addition, a map based on time units assumes that chromosome transfer proceeds at a constant rate. These disadvantages are overcome when the mapping is done on the basis of recombination frequency.

In order to determine the sequence of a given group of markers it is necessary to make the appropriate crosses involving at least three loci. For example interrupted mating between *Hfr* and F^- may reveal that a given locus *c* is a distal one and that the locus determining streptomycin resistance or sensitivity is still more distal. Two other loci, *a* and *b*, may be more proximal, but their sequence with respect to *c* cannot be easily determined by interrupting mating because they are too close together. Thus, the sequence could be either *a b c* or *b a c*. If reciprocal crosses are made in which the + + + genotype is selected, the order can be determined. Crosses I and II are shown in Fig. 8-5.

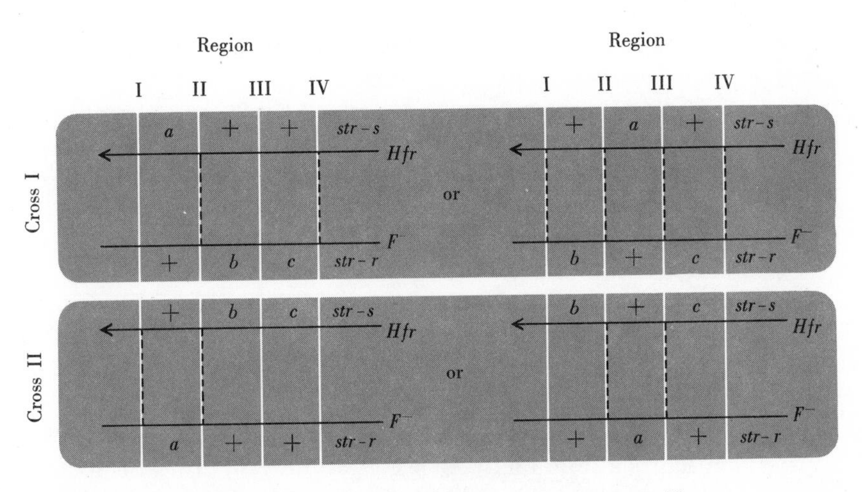

Fig. 8-5 *Crosses I and II shown diagrammatically.*

After the *Hfr* and F^- cells have been mixed they are plated to a minimal medium containing streptomycin on which the only surviving genotype will be + + + *str-r*. The diagrams shown in Fig. 8-5 reveal the way in which this recombinant would arise in each cross. You will note that if the order is *a b c* the recombinant genotype + + + is recovered in both crosses by the occurrence of two recombinational events; if the order is *b a c*, however, four of these events are necessary in Cross I and two in Cross II. If we assume that the occurrence of four recombinational events is less frequent than the occurrence of two, the order is *a b c* rather than *b a c*. This method of genetic analysis has been used in *E. coli* to determine both the sequence of genes in the chromosome and their distance apart in recombination map units. Further descriptions of the method and examples are given by Jacob and Wollman and by Hayes (see Further Reading).

TRANSDUCTION

Bacterial transduction, another way for gene recombination to occur in bacteria, was demonstrated in 1952 by N. Zinder and J. Lederberg. Some of the genetic characteristics of one bacterium can be transferred to another bacterium by means of a "third party." In transduction, this third party is a bacteriophage that is associated first with the genetic material of the donor's bacterium and later transfers part of the donor's genetic material to the recipient bacterium. Two distinct types of transduction have been recognized in bacteria. In one of them, called *generalized transduction*, any given segment of the donor's chromosome can be transmitted to the recipient cell via the phage. The other type, *localized transduction*, is limited to the transfer of specific segments of the donor's chromosome.

When bacteria are infected with *virulent* phages, the phages reproduce within the bacteria and ultimately destroy them by lysis. There is, however, a different kind of phage called a *temperate* phage. After infection these phages can either destroy their host, as do virulent phages, or they can establish a stable association with the host. In this stable association the phages are maintained within the bacterium in nonpathogenic forms called *prophages*. The capacity of the host bacteria to allow prophages to be reproduced and maintained within them is an hereditary one. Strains of bacteria having this capacity are called *lysogenic*.

A culture of lysogenic bacteria does, however, contain free phages. These arise from the lysis of a small percent of host cells. The factors that bring about this apparently spontaneous phage production within lysogenic strains of bacteria are poorly understood. It is, however, possible to induce the release of free phages by irradiating the lysogenic bacteria with ultraviolet light. The mechanism for this release may be the same as the crossover mechanism described above for the release of the *F* factor (Fig. 8-2). Usually, intact phage DNA is released, and normal phages are produced. However, on occasion, the crossover occurs in a position so as to release defective phages. Their DNA contains a portion of the host DNA, but a portion of their own DNA is missing. The missing portion remains with the host chromosome. Thus, a mixture of intact and defective phages is produced.

When this phage mixture is used to infect bacteria that differ in genotype, a small number of these bacteria will produce progeny that possess the genotype of the original bacterial host. In other words, as in conjugation in *E. coli* and in bacterial transformation, bacteria of one genotype can act as a donor strain while bacteria of a different genotype are the recipient strain. In transduction, however, the transfer of genetic material is mediated by defective phages that carry a small segment of donor DNA. This segment of DNA becomes incorporated into the recipient chromosome by the postulated crossover mechanism shown in Figure 8-2.

The fact that only a single gene is usually transduced in this manner indicates that the phages carry only a small segment of the donor DNA. There are certain instances, however, when two genes are transduced, and thus it is suggested that they may be located in the same segment of DNA, in other words, that they may be closely linked. The occurrence of such linked generalized transduction has made it possible to construct genetic maps for bacteria such as *Salmonella.*

Localized transduction is exemplified by the phage λ (lambda) which can transfer only the galactose locus (*gal*) of *E. coli* from donor to recipient cells. Experiments with this phage have led to information regarding the relationship between prophage and the bacterial chromosome. It was stated earlier that the ability of a host bacterium to allow prophages to be produced and maintained is an hereditary property. Evidence has come from crosses between strains that are lysogenic and nonlysogenic for λ. These crosses showed that the property of lysogeny is inherited and that the locus determining lysogeny is closely linked to the *gal* locus. Further, it was shown that the property of lysogeny is not the consequence of the action of a gene in the bacterial chromosome; rather, it is the prophage itself that determines this property. The prophage occupies a specific site in the bacterial chromosome: one very close to the *gal* locus. If gal^+ λ^+ cells are treated with UV so as to induce phage formation, about one phage in 10^6 is capable of transferring the gal^+ character to gal^- bacteria, and these bacteria also become λ^+. No other genetic characters are transduced.

There are many interesting and important properties of λ that are beyond the scope of this text. The interested reader is referred to the books by Hayes and by Jacob and Wollman.

We have discussed both generalized transduction and localized transduction. It is important to note that *Salmonella* and *E. coli* are also capable of genetic recombination either by conjugation or transduction, and further, it is possible to obtain conjugation and genetic transfer between the two species. We see, therefore, that two kinds of nonmeiotic transmission of the genotype have been achieved by these organisms. Certain aspects of the two modes of genetic transmission seem to be similar. For example, both processes involve a one-way transfer of genetic material. This is brought about during conjugation by a mechanism that injects a segment of donor chromosome into a recipient cell. In transduction the step of transfer is brought about by bacteriophages. These phages, adsorbed onto the surface of a bacterium, inject their genetic material, which in turn pairs with that of the recipient bacterium. This step may be likened to the transfer step in *E. coli* conjugation. Then by a process that may be closely related to crossing over, the donor segment becomes integrated into the bacterial chromosome, and finally the genetic change is expressed following division of the recipient bacterium.

GENETIC RECOMBINATION IN BACTERIOPHAGE

The mode of transmission of the genetic material of bacteriophages differs from any we have seen so far. An experiment described in Chapter 1 demonstrated that DNA is the genetic material of the bacteriophage. The result of the experiment showed that after the adsorption of the bacteriophages to the surface of the host cell, the bacteriophage DNA (but not its protein) is injected into the bacterial cell. Let us now follow the process of bacteriophage reproduction more closely and, in particular, examine the mode of transmission of certain genetic characteristics of one type of virus that infects *E. coli*.

A few minutes after the injection of phage DNA into the host cell, copies of the phage DNA begin to be synthesized. These constitute what is known as a pool of *vegetative* phages. The vegetative phages reproduce logarithmically, in the manner of a population of bacteria. The concentration of new DNA in the pool reaches a critical concentration about ten minutes after infection, and the phage DNA starts to be withdrawn from the pool and become coated with protein. This step leads to the production of *mature* phages. As the maturation process proceeds, new phage DNA is synthesized and added to the pool at the same rate as mature phages are formed and *withdrawn* from the pool. The size of the pool, therefore, remains constant. Both the synthesis of phage DNA and the maturation of phages continue until the bacterial cell is disrupted by lysis and the mature phages are released. These phages are identical to the original infecting phages if the infecting phages are of only one genetic type. If, however, the infection is a mixed one, that is, if the bacterial host is simultaneously infected with two different types, the progeny obtained are either like one or the other "parental" phages or they are recombinant types. These recombinant types arise by a "mating" process that occurs among the vegetative phages. There are numerous opportunities for the recombination of the genetic material of the phages, and it is the origin of the recombinant types from such phage crosses that is of interest to us here.

When bacteria are evenly spread on the surface of medium contained in a petri dish, their growth results in a uniform, dense carpet of cells. If phages are present, they will infect the bacterial cells and lyse them to release more phages. Clear or translucent areas will appear in the carpet where bacteria have been lysed. These clear areas, called *plaques*, contain the mature phages released after infection. The bacteriophage T_4, which infects *E. coli*, produces plaques with a characteristic morphology. Wild-type T_4, for example, produces small plaques on *E. coli* strain *B*. A population of wild-type T_4 will, however, give rise to an occasional plaque that is larger and more distinct on *E. coli B*. The phages producing these plaques of different morphology are mutant types known as *r* mutants. When they are isolated and used to infect *E. coli* strain *B* they reproduce true to type; thus plaque morphology is a recognizable phenotypic characteristic of phage. In addition to a characteristic plaque morphology, wild-type T_4 can reproduce and lyse *E. coli* of a different strain known as *E. coli K*. The fact that many *r* mutants cannot lyse *E. coli K* is another way to distinguish them from wild-type T_4. This distinction is of particular importance when it is necessary to recognize a rare wild-type phage in a population of a given *r* mutant. If the phages are plated to *E. coli K*, the wild-type phage yields a plaque, and thus is easily detected.

An experiment can be performed in which *E. coli B* is infected simultaneously with two different *r* mutants, which we will call *r-a* and *r-b*. Neither of these *r* mutants can lyse *E. coli K*. The mature phages produced on *E. coli B* are then used to infect *E. coli K*. If plaques appear on *E. coli K* they must be caused by the presence of wild-type T_4, since neither mutant is capable of lysing *E. coli K*. These wild-type phages could arise either by the occurrence of a mutation of an *r* to wild type, or by recombination between the two *r* mutants. The rate of mutation *r* to + is generally very low, and it can be determined for the two *r* mutants by infecting *E. coli K* separately with each *r* mutant. This control experiment makes it possible to determine the number of wild-type phages that might have arisen by mutation in the mutant strain. The number depends upon the nature of the *r* mutant. For some *r* mutants there will be as few as one wild-type phage for every 10^8 mutant phages; for others, a different number is obtained. The mixed infections of two different *r* mutants, however, usually yield more wild-type plaques than can be accounted for by mutation. Therefore, their origin must be through recombination, and the number of plaques found on *E. coli K* is a measure of the frequency of recombination.

It is found, for example, that *r-a* and *r-b* recombine to give wild type at a frequency of five wild type for every 100 *r-a* and *r-b*. Other *r* mutants give higher frequencies of wild type, while still others give a frequency of recombinants as low as 1×10^{-6}. This exceedingly low frequency of recombination can be recognized because the rare occurrence of a single wild-type phage can be detected. This ease of detection results from the fact that the phage is able to produce progeny on *E. coli K*, and thus a plaque is formed. The ability to detect a rare recombinational event provides for a very high degree of genetic resolving power—much higher, for example, than is practical with nucleate organisms. Thus the genetic map of the *r* mutants in phage can be resolved to very small distances. In fact, these distances are approximately equal to the distance between only a few nucleotide pairs of a phage DNA molecule.

S. Benzer has made an extensive analysis of the *r*II mutants of phage T_4. From 1000 to 1500 different *r*II mutants have been arranged in a linear sequence within one segment of the phage chromosome. The two most distant mutants give a recombination frequency of about 10 percent. The closest mutants give approximately 0.02 percent recombination, and there are many mutants between which recombination has not been detected. All of the mutants in the *r*II region are concerned with a similar function. That is, they fail to produce plaques on *E. coli* strain *K*. This *inactive* phenotype is to be contrasted with the *active* phenotype of wild-type phage, which does produce plaques on this strain of *E. coli*.

Thus far we have been studying the phenomenon of *recombination* which occurs when mutant phages simultaneously infect *E. coli* strain *B*; the recombinant phages are then detected by allowing the progeny to infect *E. coli* strain *K*. A dilute suspension of the phage progeny must be used in this second infection so that a given plaque formed on the carpet of K cells contains the progeny of only *one* recombinant phage which has originally infected one *K* cell. If this precaution is not taken, more than one phage may infect a *K* cell and a second phenomenon, that of *complementation*, may occur.

Consider first the result of an experiment in which *K* cells are simultaneously infected with a mixture of wild-type phages and *r* mutant phages. Plaques are produced that contain not only wild-type phages but the mutant phages as well. This is possible because the

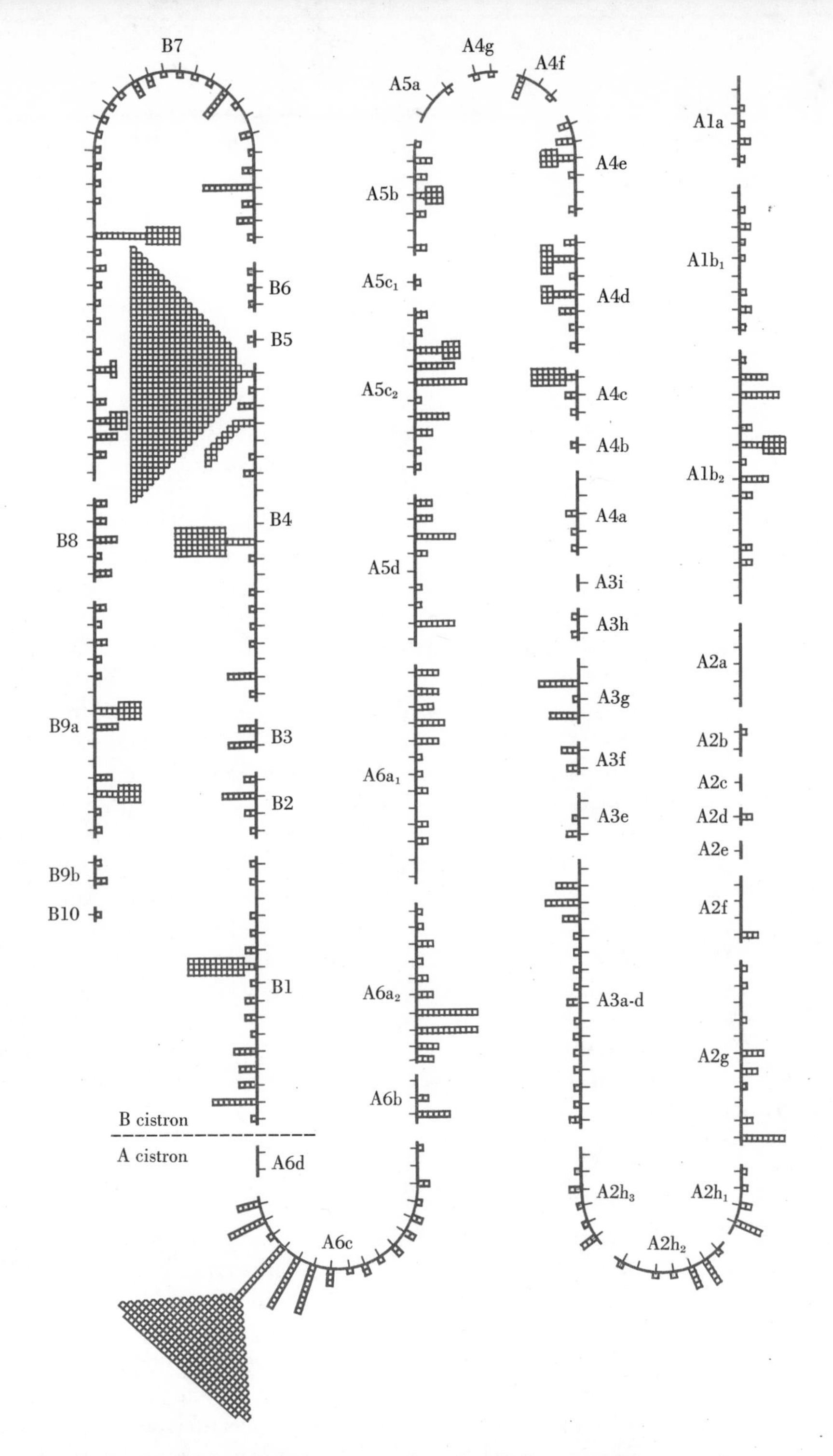

Fig. 8-6 *The order of mutations within the* r*IIA and* r*IIB genes of* T_4*. Each square identifies the location in the chromosome of an independently occurring mutation. Where more than one mutation occurs, the number of squares indicates the number of mutations. (After Benzer,* Proceedings of the National Academy of Sciences, *Vol. 47, p. 410, 1961.)*

wild-type phages are able to carry out the functions necessary for lysis of the *K* cell, and thus the mutant phage is able to reproduce even though no recombination has occurred.

Now consider the result of a mixed infection of two different rII mutants in a *K* cell. If one mutant is able to supply the function that is missing in the other and vice versa, lysis can occur and plaques are recovered that contain both kinds of mutant phages as well as any recombinant phages. If the two mutants cannot *complement* one another in function, the only plaques that are formed are those which contain wild-type recombinant phages alone.

Benzer applied such a *cis-trans* test (see Chapter 6) on various pairs of mutants, and found that certain pairs could complement one another while other pairs could not. He was thus able to divide the rII region into two cistrons called *A* and *B*. Mutants within cistron *A*, or within cistron *B*, are deficient in the same function and therefore noncomplementary, but any mutant from cistron *A* will complement with any mutant in cistron *B*. Two different functions are therefore needed to perform the lysis of *K* cells; any of a number of defects in either cistron will result in a mutant phage. The genetic map of the *r* IIA and *r* IIB genes is shown in Fig. 8-6.

There is a variety of phage mutants that affect, in addition to plaque morphology, the range of bacterial hosts that can be attacked by the phage, the color of plaques, the turbidity of plaques, and the temperature at which plaques are formed. All of these mutant types have been used to obtain genetic maps of phage. Thus the phage, which consists primarily of DNA and a protein coat, has its genetic material organized in a fashion not dissimilar to that of the nucleate and nonnucleate organisms we have examined. The genetic loci are arranged in a linear order, and recombination can occur within a gene as well as between different gene loci.

BACTERIAL TRANSFORMATION

Transformation entails the exposure of bacteria of one genotype to purified DNA derived from bacteria of a different genotype. In order for transformation to occur, the recipient cells must be in a *competent* state. The nature of this state and the factors that determine it have not been fully defined. However, the competent state most likely represents a physiological condition that permits the uptake or incorporation of the donor DNA. After the DNA is incorporated, it becomes integrated with the chromosome of the recipient by recombination.

The DNA incorporated and integrated by a recipient bacterium represents a small percentage of the total DNA of any given donor bacterium. Transformation can occur for a single gene and conceivably for more than one gene represented in a single segment of donor DNA. It can also occur for two or more genes, each of which is in a different segment of donor DNA. Consider the simultaneous transformation of two different genes. If it occurs at a frequency less than or equal to the product of the frequency of the single transformation of each gene, the simultaneous, or double, transformations are independent events. This is interpreted to mean that the two genes are in different segments of DNA that have been incorporated and integrated independently of each other. If, however, the frequency of the double transformation is greater than expected for two independent transformations, the two genes are considered to be in the same segment of DNA. Accordingly, when a double transformation arises as the occurrence of two independent events the two genes are said to be unlinked, and when the double transformation is the consequence of a single event the two genes are said to be linked. For example, in pneumococcus the genes for mannitol utilization (mtl^+) and streptomycin resistance (*str-r*) are linked. This linkage has been demonstrated by showing that when the donor DNA is obtained from mtl^+ *str-r* cells, the frequency of doubly transformed mtl^- *str-s* recipient cells exceeds that of the product of the two single transformations for mtl^+ and *str-r* in experiments where the donor DNA is obtained from cells that are mtl^+ *str-s* and mtl^- *str-r*.

The linkage relationship between several genes of *Bacillis subtilis* has been determined in a most ingenious fashion by N. Sueoka and his colleagues by taking advantage of the fact that the chromosome replicates in a sequential fashion starting from a point of origin (Fig. 8-7). The order of several different genetic markers was obtained by using for transformation DNA that was isolated from cells at various times

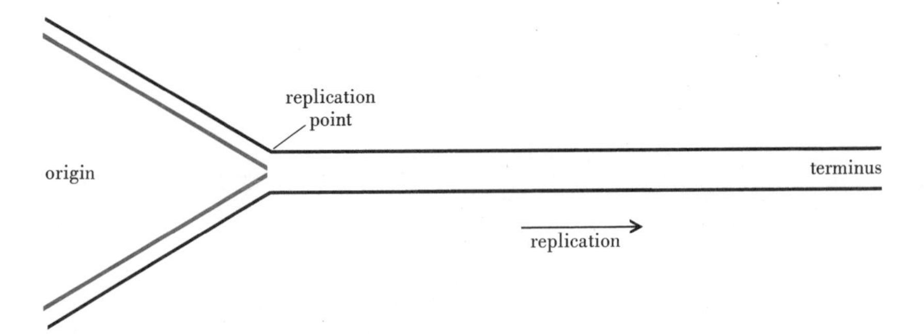

Fig. 8-7 *Model for the replication of the chromosome of* Bacillus subtilis. *Replication starts at one end, the origin, and proceeds to the other end, the terminus. Both strands are copied and the black line represents the old strand and the color line the newly synthesized strand (after Yoshikawa and Sueoka,* Proceedings of the National Academy of Sciences, *Vol. 49, 1963).*

during the exponential growth phase. The DNA so isolated was in different stages of its sequental replication. For DNA isolated early during replication, it would be expected that markers closer to the origin would be represented more frequently than markers closer to the terminus. Thus, the frequency of transformation for the markers present in the replicated portion of the DNA should be higher than for those in the portion of the DNA that has not yet replicated. A genetic map obtained by measuring the frequency of transformation using donor DNA that was isolated at different times during its replication is shown in Fig. 8-8. The position of each marker is based upon the frequency with which a given marker was recovered relative to the terminal marker, *met.*

Clearly, the phenomenon of transformation is of fundamental significance to genetics, since we see that it occurs at the level of the genetic material itself. It might be thought, however, that transformation is only a laboratory device and that it has no significance outside of providing a model situation for the investigation of the direct role that DNA plays in the transfer of genetic characteristics. This might well be the case, but recall the experiments of Griffith described in Chapter 1. It is true that these experiments were not wholly "natural," since Griffith used heat-killed cells as the source of transforming DNA; however, more recent experiments have demonstrated transformation carried out in a culture containing two different strains of living bacteria. One was resistant to a sulfa drug and a toxic substance called amethopterin, whereas the other was resistant to streptomycin and another

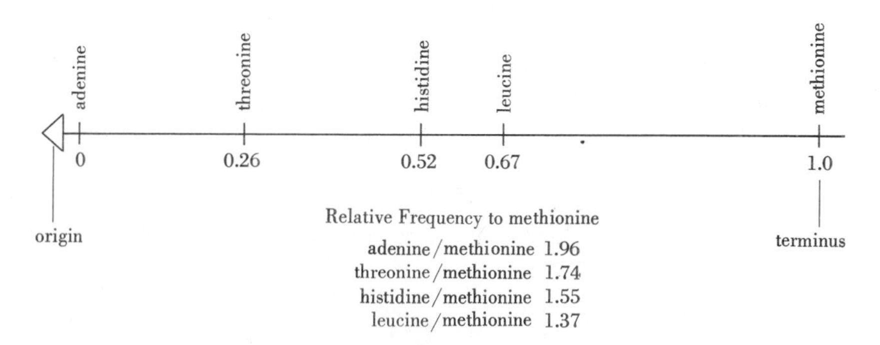

Fig. 8-8 *Map of the* Bacillus subtilis *chromosome. The position of each marker is determined relative to the terminal marker,* met. *The figures given below the map are the relative frequencies, with respect to* met, *for the transformation of the four markers (after Yoshikawa and Sueoka,* Proceedings of the National Academy of Sciences, *Vol. 49, 1963).*

antibiotic micrococin. The two strains were mixed and allowed to grow through several divisions. Then the cells were harvested and tested for resistance to the various substances. The results, which may seem obvious, were colonies that were resistant to all four substances. A similar experiment was performed in which cells of one genotype were grown in a medium that had previously contained cells of a different genotype. When this culture was tested, it was found to contain the recombinant type of bacteria. This experiment rules out the possibility of conjugation between donor and recipient bacteria. Both experiments reveal that in cultures under laboratory conditions lysis of cells, most likely at their death, releases into the medium DNA, which can then be taken up by living cells. The recombinant bacteria did not arise when DNAse was present in the medium indicating that free DNA was responsible for the transformation.

If these cells differ in genotype, recombinants can then be formed. Coupled with Griffith's work, these experiments suggest that transformation may take place not just in laboratory cultures, but among bacteria that infect some appropriate host. It may be possible that *in vivo* bacterial recombination, following conjugation or transduction, can also occur within the normal host animals, as can phage recombination. Thus, the various types of nonmeiotic transfer of the genetic material might occur in nature under appropriate environmental conditions.

FURTHER READING

Adelberg, E. A. (ed.), 1960. *Papers on bacterial genetics.* Boston: Little, Brown & Company.

Hayes, W., 1964. *The genetics of bacteria and their viruses.* Oxford: Basil Blackwell & Mott, Ltd.

Jacob, F., and E. L. Wollman, 1961. *Sexuality and the genetics of bacteria.* New York: Academic Press, Inc.

Stent, G. S. (ed.), 1960. *Papers on bacterial viruses.* Boston: Little, Brown & Company.

Watson, J. D., 1965. *The molecular biology of the gene.* New York: W. A. Benjamin, Inc.

part III

THE ACTION OF THE GENETIC MATERIAL

chapter 9

Mutation

A consideration of the phenomenon of mutation brings us to the third, and final, part of this book—namely, the question of the action of the genetic material. Gene mutation implies that the genetic material can undergo some sort of change that results in the production of an altered phenotype.

DEFINITION OF MUTATION

A white-eyed fly may be found in a culture of wild-type *Drosophila*; a culture of *Neurospora* may have an occasional colony that has white rather than colored conidia; and, a strain of *E. coli*, sensitive to the bacteriophage T_1, may give rise to cells that are resistant to this bacteriophage. These three examples illustrate mutant types that arise spontaneously within "pure lines" of these organisms. When the genetics of these mutants is investigated it is found that each mutation is transmitted according to the patterns of inheritance already described for the organisms in question.

Mutation can be defined, therefore, as an event that gives rise to a heritable alteration in the genotype. Mutations are divided into two major classes: those involving a change in the structure or number of chromosomes and those involving a single gene.

CHROMOSOMAL MUTATIONS

There are two major classes of chromosomal aberrations: first, changes in chromosome number; and second, those changes in the chromosome structure that result in alterations of gene order or number. Examples and descriptions of chromosomal mutations are shown in Fig. 9-1.

Drosophila and certain other insects possess large and distinct chromosomes in the nuclei of their salivary gland cells. These so-called *giant* chromosomes are particularly favorable for cytological observations of chromosome mutations, for, in addition to their relatively large size, each chromosome pair is readily recognized according to a specific pattern of bands along its length. This pattern is so specific that chromosome mutations that cause deletion, duplication, or rearrangement of a single band can be detected.

A *deficiency* or *deletion* is illustrated in part A of Fig. 9-1. The diagrams in *a* through *c* illustrate how the deficiency may come about through the occurence of two breaks in the chromosome followed by the reunion of the broken ends accompanied by the loss of a segment of the chromosome. Pairing between a normal and a deficient chromosome is seen diagrammatically in *d* whereas the actual configuration seen in the *Drosophila* salivary gland chromosomes is shown in *e*.

A class of chromosome mutation known as a *duplication* is illustrated in part B of Fig. 9-1. *Bar* eye in *Drosophila* is a duplication. When the segment identified as 16A in the X chromosome is duplicated, the flies are *Bar* in phenotype rather than wild type. When a second duplication occurs, the flies show a more extreme form of *Bar* eye known as *double Bar*.

A frequently encountered chromosome mutation in *Drosophila* is one in which the sequence of genes in a chromosome becomes inverted. This change in sequence is known as *inversion*, and it can be identified genetically by the alteration of the genetic map of the chromosome. However, it can also be detected cytologically as shown in part C of Fig. 9-1. A normal and an inverted chromosome are shown diagrammatically in *a* and *b*. When they pair (*c* or *d*) it can be seen

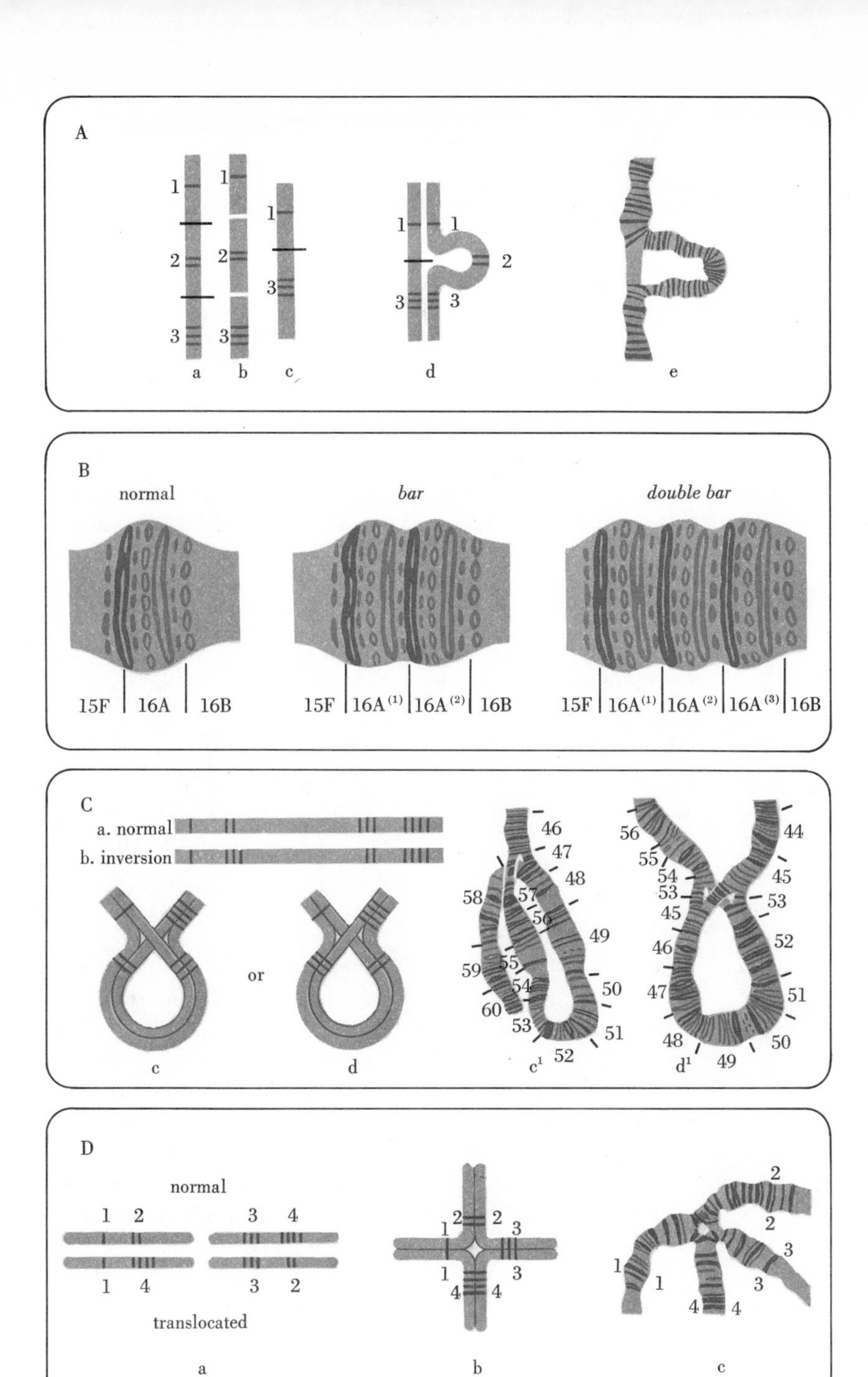

Fig. 9-1 *Four of the several different kinds of chromosomal mutations as exemplified in* Drosophila. *(After Altenburg,* Genetics. *New York: Holt, Rinehart and Winston, 1957.)*

that one or the other of the two chromosomes forms a loop so that the homologous segments of each chromosome become aligned. The actual pairing configuration in the salivary gland chromosomes is shown in c' and d'.

A chromosome mutation known as a *translocation* occurs when a segment from one chromosome becomes attached to a nonhomologous chromosome. If two nonhomologous chromosomes exchange segments, it is known as a *reciprocal translocation.* This is shown in part D. Two normal, nonhomologous chromosomes and two translocated chromosomes are shown in *a*. The pairing configuration of normal chromosomes with those having a reciprocal translocation is shown diagrammatically in *b*. As in the case of an inversion, the pairing configuration reflects the positioning of the homologous segments of the chromosomes. An actual reciprocal translocation is shown in *c*.

More specific information concerning chromosome mutations can be found in the book by Swanson referred to at the end of this chapter.

These mutations reflect gross changes in the organization of the genetic material, whereas gene mutations represent changes that have occurred at the level of one to a few nucleotides within the DNA molecule and below the resolution of the electron microscope.

GENE MUTATION

In *Drosophila* the white-eye phenotype results from the inability of the organism to syntheisze the red and brown eye pigments characteristic of wild-type flies. Clearly, the mutational event $+ \rightarrow w$ results in a change in the genotype such that the mutant allele w lacks, within the DNA, the genetic information necessary for production of these pigments, and it thus represents the loss of a specific genetic function. There are many mutations in *Drosophila* affecting the synthesis of eye pigments. Two of them, *st (scarlet)* and *bw (brown)*, are recessive and are in the third and second chromosomes, respectively. When they are crossed

$$\left(\frac{st\ +}{st\ +} \times \frac{+\ bw}{+\ bw}\right)$$

the F_1 heterozygotes have a wild phenotype. Inbreeding the F_1 leads to a phenotypic ratio of 9 *wild type* : 3 *scarlet* : 3 *brown* : 1 *white* which is the ratio expected from a two-factor cross. The wild-type eye color results from the presence of two pigments, brown and red. The mutation $+ \rightarrow st$ results in the loss of ability of *st/st* flies to synthesize brown pigment; hence they have only the red pigment. The mutation $+ \rightarrow bw$ leads to the loss of ability of *bw/bw* flies to synthesize the red pigment, and thus these flies have only the brown pigment. Clearly, the $\frac{st\ bw}{st\ bw}$ flies can synthesize neither pigment and are, therefore, white-eyed. The many eye-color mutations of *Drosophila* represent, in varying degrees, changes in the ability to synthesize the two pigments.

These eye-color mutants are a few of the large class of so-called *visible* mutations whose phenotypic effects are seen as an alteration in the organism's morphology. Visible mutations may be either recessive, as in *white* eye, or dominant, as in the case of *Bar* eye. They are known in many different kinds of organisms; for example, *taillessness* in mice and Manx cats, *albinism* in man, and *white* flower color in sweet peas, to name but a very few. Many visible mutations represent the loss of some essential function or structure. Other mutations can lead to the addition of certain structures, or to the extreme modification of others. For instance, in *Drosophila* the mutation *proboscipedia* causes the proboscis to develop into a leglike structure, and in man polydactyly is a mutation with a phenotypic effect expressed in the form of extra fingers. Visible mutations have effects that run the gamut from what appear to be relatively simple alterations in phenotype to those that are major morphological departures from wild type.

A second class of gene mutations affects specific biochemical processes. These have been studied extensively in a variety of microorganisms and are called *biochemical mutations*. They often represent the loss of a specific biochemical function, since the mutant organism has lost the ability to synthesize an essential metabolite such as an amino acid or vitamin. There are also biochemical mutations that result in only a partial loss of a synthetic step, or the inability to carry out the step under certain external conditions such as temperature, pH, or the composition of the growth medium. As we shall see in this and the next chapter, a biochemical mutation can be ascribed to the loss or modification of the organism's ability to make a specific protein, most often an enzyme necessary for the normal function of some essential process. Such a loss or modification is undoubtedly the underlying cause of most mutations.

Biochemical mutations are lethal unless the organism is provided with the metabolite it is not able to synthesize. There is a third class of mutations, widely studied in *Drosophila* and other higher organisms, known as *lethal mutations*. These mutations cause the death of the organism and many represent the loss or alteration of a function essential during the embryology of the organism. If the biochemical function that is lost in these mutants could be repaired, they would no longer be classified as lethals, but at present the nature of lethal mutations in higher organisms is not well enough understood to permit their correction.

SPONTANEOUS MUTATION Organic evolution is a process that either preserves or rejects different genotypes. These differences in genotype result ultimately from the accumulation of different mutant genes within natural populations of organisms. The process of mutation is the only means known whereby new types of genes, which serve as the raw material for evolution, can arise. Spon-

Table 9-1 **Spontaneous Mutation Rates in Different Organisms**

Organism	*Character*	*Rate*	*Units*
BACTERIOPHAGE:	lysis inhibition, $r \rightarrow r^+$	1×10^{-8}	per gene[a]
T_2	host range, $h^+ \rightarrow h$	3×10^{-9}	per replication
BACTERIA:	lactose fermentation, $lac^- \rightarrow lac^+$	2×10^{-7}	
Escherichia	phage T_1 sensitivity, T_1-$s \rightarrow T_1$-r	2×10^{-8}	
coli	histidine requirement, $his^- \rightarrow his^+$	4×10^{-8}	
	$his^+ \rightarrow his^-$	2×10^{-6}	per
	streptomycin sensitivity,		cell
	str-s → *str-d*	1×10^{-9}	per
	str-d → *str-s*	1×10^{-8}	division
ALGAE:	streptomycin sensitivity,		
Chlamydomonas reinhardi	*str-s* → *str-r*	1×10^{-6}	
FUNGI:	inositol requirement, $inos^- \rightarrow inos^+$	8×10^{-8}	mutant
Neurospora	adenine requirement, $ade^- \rightarrow ade^+$	4×10^{-8}	frequency
crassa			among asexual spores
CORN:	shrunken seeds, $Sh \rightarrow sh$	1×10^{-5}	
Zea mays	purple, $P \rightarrow p$	1×10^{-6}	
FRUIT FLY:	yellow body, $Y \rightarrow y$, in males	1×10^{-4}	mutant
Drosophila	$Y \rightarrow y$, in females	1×10^{-5}	frequency
melanogaster	white eye, $W \rightarrow w$	4×10^{-5}	per
	brown eye, $Bw \rightarrow bw$	3×10^{-5}	gamete
MOUSE:	piebald coat color, $S \rightarrow s$	3×10^{-5}	per
Mus musculus	dilute coat color, $D \rightarrow d$	3×10^{-5}	sexual
MAN:	normal → hemophilic	3×10^{-5}	generation
Homo sapiens	normal → albino	3×10^{-5}	
HUMAN BONE	normal → 8-azoguanine resistant	7×10^{-4}	
marrow cells	normal → 8-azoguanosine resistant	1×10^{-6}	per cell
in tissue culture			per division

[a] Correction of the other mutation rates in this table to a per-gene basis would not change their order of magnitude.

From R. Sager and F. J. Ryan, *Cell Heredity*. New York: John Wiley & Sons, Inc., 1961.

taneous mutation rates have been measured in a variety of organisms, and, as the figures in Table 9-1 show, they vary not only from organism to organism, but also from gene to gene within the same organism. All spontaneous mutations occur at a low rate. However, since evolution in higher organisms is viewed in terms of geological time, the relatively slow rate of spontaneous mutation per generation becomes very significant in a vastly expanded time scale. (For a detailed discussion, see *Evolution*, in this series.) In populations of microorganisms, generation time is relatively short, and the population size can be enormous. Hence mutation operates as an evolutionary force over a shorter period of time than it does for populations of higher organisms.

A knowledge of the existence of spontaneous mutations antedates modern genetics by many years, since the sporadic occurrence of "sports" was recognized among domestic animals and plants by breeders in many parts of the world. In 1910, T. H. Morgan used white-eyed flies to establish the existence of sex-linked inheritance. These flies were the descendants of a single white-eyed male that appeared spontaneously in a culture of wild-type flies. Furthermore, the many different mutant types that Morgan and his students used in their experiments arose spontaneously. That is, they were not induced by artificial means through the use of *mutagenic agents*. In fact, it was not until almost two decades later that it was discovered that mutations could be induced.

THE INDUCTION OF MUTATIONS

In 1927 H. J. Muller demonstrated that x rays could induce mutations in *Drosophila*. Muller's findings were followed the next year by those of L. J. Stadler, who used x rays to induce mutations in barley. Their discovery is among the most significant contributions to genetics, for with x rays it became possible to obtain large numbers of mutant genes in many organisms. These mutants are essential for the investigation of the transmission of the genetic material, for they are the tags by which the genetic material can be followed from one generation to the next. Moreover, the Muller-Stadler discovery initiated the investigation of the nature of the mutation process itself and the relationship between mutation and the nature of the gene.

X rays, as well as alpha, beta, and gamma rays, belong to the class of radiations known as *ionizing* radiations. All are mutagenic. Experimental studies on the induction of mutations with ionizing radiations have been carried out with a variety of organisms, and their results have led to several important conclusions. First, ionizing radiations can cause both gene and chromosome mutations. Second, the rate of occurrence of mutations is a function of the dose of x rays that the organism receives. This function, in the case of gene mutations, appears to be a linear one, as seen in the case of the induction of sex-linked lethal mutations in *Drosophila* (Fig. 9-2). Third, the mutagenic effect of ionizing radiations is cumulative; the rate of mutation is a function of the total amount of radiation which the organism has received. For example, an acute x-ray dose of 6000 roentgens (the roentgen is a measure of the amount of ionization produced by the x rays) results in the production of 12 percent sex-linked lethal mutations in *Drosophila*. If this amount of irradiation is given in three 2000-roentgen doses, spaced eight hours apart, the same percentage of sex-linked lethals is usually obtained. Therefore, relatively low doses of chronic irradiation in *Drosophila* can lead to the same genetic effect as one acute dose. In mice, however, chronic irradiation with low doses of x rays is less mutagenic than an acute dose of the equivalent amount of irradiation.

This fact is related to the question of the genetic effects of the low, chronic dose of irradiation to which the entire human population is exposed through the natural radiation of the earth and cosmic rays and through fallout radiation from the explosion of nuclear weapons. There is at present no unequivocal evidence regarding the extent of the genetic effects of this chronic dose. However, it would seem safe to assume that the presence of fallout radiation, and certainly of any increase in this source of radiation, poses a serious potential hazard for future human generations. Most new mutations in man, and other organisms as well, are deleterious; their phenotypic effects are, in general, less favorable than those of their wild-type alleles. The long history of evolution has preserved the favorable alleles and rejected those that are less favorable. As a consequence, a newly arisen mutation has only a small probability of producing a favorable phenotype under normal environmental conditions. Therefore, any increase in the amount of radiation to which we are exposed may produce new mutations whose debilitating effects will be borne by future generations.

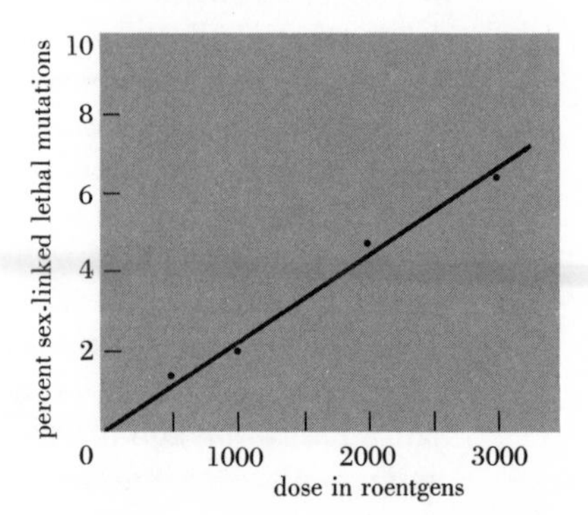

Fig. 9-2 *The relation between x-ray dose and the frequency of sex-linked lethal recessive mutations in* Drosophila.

The manner in which ionizing radiations produce mutations is not fully understood. One possibility is that they act directly on the genetic material. This direct action of the radiation can be likened to a bullet hitting a target. A quantum of radiant energy hits a gene and alters or destroys a portion of the genetic material. The linear relationship between the dose of irradiation and the number of mutations, as well as the cumulative mutagenic effects of x rays, seem to be in accord with the direct action of the ionizing radiation. A second possibility is that x-ray-induced mutations result from chemical changes that the ionizing radiation produces within a cell. The irradiation may produce highly reactive compounds of short life span which then act upon the genetic material, causing it to mutate. Thus, the genetic effect of x rays would be indirect.

Mutations can be induced with a variety of other agents. Among them are ultraviolet light and various chemicals. Certain mutagens act upon DNA in a specific manner, and it has thus become possible to interpret their action in molecular terms. These mutagens will be discussed later in the chapter.

MUTATION AND THE LOSS OF PROTEIN FUNCTION

Once mutant organisms had been obtained and methods for inducing mutations had been developed, it became possible to investigate the molecular nature of the alterations that had occurred as a result of mutation. The study of biochemical mutations has provided much of the needed information.

One of the earliest investigations of biochemical mutations concerned the metabolic disease of man known as *alkaptonuria*. Approximately one person in every million has this disease, and analysis of pedigrees of families in which one or more individuals are alkaptonuric reveals that the metabolic disorder is inherited as a recessive trait. The affected individuals, (*a*/*a*), in addition to having arthritic handicaps, are unable to metabolize a substance called homogentisic acid. Normal individuals convert homogentisic acid into acetoacetic acid, which in turn is excreted in the urine (Fig. 9-3). A single biochemical reaction is missing in alkaptonurics, for they cannot convert homogentisic acid into acetoacetic acid. They excrete homogentisic acid in their urine, and the homogentisic acid oxidizes upon exposure to air, causing the urine to become black. The chemical reaction by which homogentisic acid is converted to acetoacetic acid requires the presence of an enzyme that is either inactive or absent in alkaptonurics.

In contrast to the mutation causing alkaptonuria, some biochemical mutations exhibit a variety of phenotypic effects. In the inherited human disorder known as *phenylketonuria* the affected individuals,

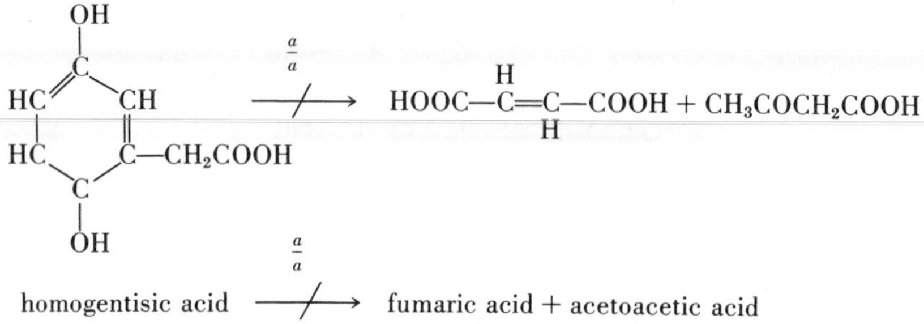

$$\text{(homogentisic acid ring: } C(OH)\text{, HC, HC, CH, } C{-}CH_2COOH\text{, } C(OH)) \xrightarrow[\not\;]{a/a} HOOC{-}\underset{}{\overset{H}{C}}{=}\underset{H}{C}{-}COOH + CH_3COCH_2COOH$$

$$\text{homogentisic acid} \xrightarrow[\not\;]{a/a} \text{fumaric acid} + \text{acetoacetic acid}$$

Fig. 9-3 *Conversion of homogentisic acid into acetoacetic acid.*

(p/p), excrete phenylpyruvic acid in their urine, whereas normal individuals do not. In addition, those affected are mentally retarded, and the amount of their hair and skin pigmentation is reduced. As in the case of alkaptonuria, phenylketonuria may be attributed either to the absence or alteration of a single enzyme. The reaction absent in the affected individuals is the conversion of the amino acid phenylalanine into the amino acid tyrosine. It is known that this reaction requires a single enzyme. In phenylketonurics, the phenylalanine is converted into phenylpyruvic acid instead of tyrosine (Fig. 9-4). Both examples suggest the existence of some relationship between a specific gene and a specific enzyme, indicated, in each example, by the absence of one enzymatically controlled reaction. Normal individuals possess the enzymes necessary for the two reactions considered. Those individuals who differ by either of the two single gene mutations produce no enzyme, or perhaps produce an enzyme that has been so modified as to be inactive.

Detailed knowledge of the gene-enzyme relationship stems largely from research with a number of different microorganisms, and many concepts of this relationship were derived from experiments with *Neurospora* initiated by G. W. Beadle and E. L. Tatum. In order to understand the gene-enzyme relationship we will begin by studying an example of one of the roles that genes play in the control of a *biosynthetic pathway.* In a living system the synthesis of organic molecules, such as amino acids, is called biosynthesis. Biosynthesis entails in most cases a series of stepwise biochemical reactions, each catalyzed by an enzyme. The series of reactions that leads to the synthesis of a given compound—arginine, for example—constitute a biosynthetic pathway.

A wild-type strain of *Neurospora* can synthesize all of the amino acids essential for the formation of its proteins. There are several mutant strains, however, that are unable to synthesize the amino acid

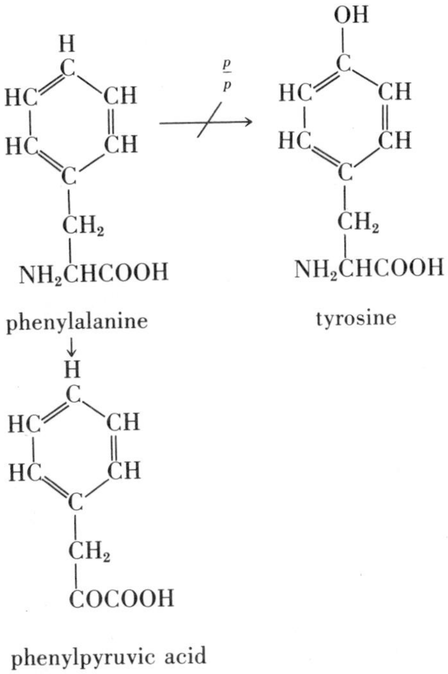

H
C
HC CH
HC CH
C
CH_2
$NH_2CHCOOH$

phenylalanine

$\xrightarrow{p/p}$ (blocked)

OH
C
HC CH
HC CH
C
CH_2
$NH_2CHCOOH$

tyrosine

↓

H
C
HC CH
HC CH
C
CH_2
COCOOH

phenylpyruvic acid

Fig. 9-4 *Breakdown of phenylalanine into tyrosine by normal individuals, and into phenylpyruvic acid by phenylketonuric individuals.*

arginine. In other words, there are several different gene loci that are responsible for the biosynthesis of arginine, and a mutation at any one of them results in a block in arginine biosynthesis. Each mutant locus segregates in a one-to-one fashion from its wild-type allele. The arginine biosynthetic pathway in *Neurospora* is known to involve reactions that go in this sequence: precursors of ornithine→ornithine→citrulline→arginine (Fig. 9-5A). Since the steps in the pathway are known, it is possible to ascertain where synthesis is blocked in each mutant strain. For example, one class of mutants will grow only when supplied with arginine; another will grow *either* on arginine or on one of its precursors, the amino acid citrulline; a third will grow not only on arginine or citrulline, but also on ornithine—another of the amino acid precursors of arginine (Fig. 9-5B). No known mutants grow on ornithine or on arginine but not on citrulline. One type of arginine-requiring mutant that can grow in the presence of ornithine is able to carry out the reactions from ornithine through citrulline to arginine. Therefore, this mutant will also grow when given either citrulline or arginine. Thus, gene mutation in this class of mutant results in a block in one of the reactions that precedes the synthesis of ornithine. Simi-

A

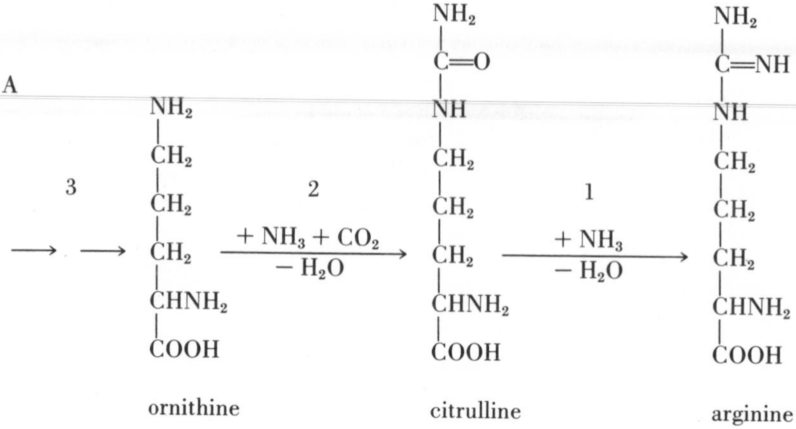

$$\xrightarrow{3}\ \longrightarrow \begin{array}{c} NH_2 \\ | \\ CH_2 \\ | \\ CH_2 \\ | \\ CH_2 \\ | \\ CHNH_2 \\ | \\ COOH \end{array} \xrightarrow[-\,H_2O]{\overset{2}{+\,NH_3\,+\,CO_2}} \begin{array}{c} NH_2 \\ | \\ C{=}O \\ | \\ NH \\ | \\ CH_2 \\ | \\ CH_2 \\ | \\ CH_2 \\ | \\ CHNH_2 \\ | \\ COOH \end{array} \xrightarrow[-\,H_2O]{\overset{1}{+\,NH_3}} \begin{array}{c} NH_2 \\ | \\ C{=}NH \\ | \\ NH \\ | \\ CH_2 \\ | \\ CH_2 \\ | \\ CH_2 \\ | \\ CHNH_2 \\ | \\ COOH \end{array}$$

ornithine | citrulline | arginine

B

arginine-requiring mutant	growth on: minimal	ornithine	citrulline	arginine	reaction blocked
1	−	−	−	+	1
2	−	−	+	+	2
3	−	+	+	+	3

Fig. 9-5 *(A) Part of the pathway of arginine in biosynthesis in* Neurospora; *(B) growth requirements of three classes of arginine-requiring mutants in* Neurospora.

larly, a mutant that can grow in the presence of citrulline or arginine, but *not* ornithine, must be unable to convert ornithine into citrulline; and accordingly, a mutant that can utilize only arginine must possess a block between citrulline and arginine. The position of each of the blocks in arginine synthesis is indicated in Fig. 9-5A and the number at each block corresponds to one of the three types of mutants shown in Fig. 9-5B.

Genetic analysis of recombination has shown that the three classes of mutant genes are nonallelic. The evidence derived from examining the blocks in arginine biosynthesis shows that each of the three mutant classes has lost one of three different biochemical reactions. Thus, there is a genetic as well as a functional distinction between them. A different enzyme is required for each of the steps in the biosynthesis of arginine. We can assume, therefore, that the functional distinction between the wild type and the three classes of arginine-requiring mutants resides either in the loss of ability to produce one of three enzymes essential for arginine biosynthesis, or in the production of an altered enzyme.

By use of the arginine-requiring mutants, it can be shown that mutants that are blocked at *different* points in the pathway of arginine synthesis form heterokaryons (see Chapter 6) that grow on minimal medium. For example, a heterokaryon that is composed of nuclei from a mutant blocked between ornithine and citrulline and a mutant blocked between citrulline and arginine grows on minimal medium. A wild-type gene in the nuclei of the former causes the formation of the enzyme that carries out the synthetic step between citrulline and arginine, whereas a wild-type gene in the nuclei of the latter causes the formation of the enzyme that carries out the step from ornithine to citrulline. Thus, the two functionally different nuclei in a common cytoplasm are able to complement each other. This complementation leads to growth on a minimal medium that is indistinguishable from that of wild type. If, however, the nuclei are derived from two mutant strains, both of which have a block between ornithine and citrulline, the heterokaryon will not grow on a minimal medium, and the two strains are noncomplementary. Thus, the two strains are unable to carry out the same synthetic step, and it is implied that the same enzyme is missing or altered in both and that the two gene loci are functionally identical.

In these examples gene mutation has resulted in an alteration of the genetic control of a single biochemical event, expressed as the loss of the organism's ability to carry out *one* of the essential steps in a biosynthetic pathway. The loss of a single enzyme-dependent step in a biosynthetic pathway has been demonstrated in mutant organisms unable to synthesize vitamins, purines, pyrimidines, and amino acids. The fact that each step requires a different protein suggests that the *function of a gene can be equated with the formation of a protein.* If this is the case, it should be possible to demonstrate that a gene mutation leads either to the lack of synthesis of a specific protein or to the synthesis of an altered protein.

MUTATION AND THE ALTERATION OF PROTEIN STRUCTURE

The *primary structure* of a protein is its sequence of amino acids. If the sequence is altered by the substitution of one amino acid for another at one site in the polypeptide, the protein may have entirely different properties. Changes in amino acid sequence have been found, for example, in human hemoglobin, in the protein of the tobacco mosaic virus (TMV), and in the enzyme tryptophan synthetase from *E. coli.* The fact that each of the altered proteins has been shown to have arisen by mutation implies that genes control protein primary structure.

The first instructive picture of the genetic control of amino acid sequence came from V. Ingram's studies of human hemoglobin. There are different forms of adult hemoglobin recognized by virtue of differences in electric charge. Normal adult hemoglobin (hemoglobin *A*) is considered wild type. The remaining adult hemoglobins are mutant forms and are related to a series of anemias, some of which are fatal. For example, sickle-cell anemia is a disease found most frequently among certain African populations, and persons suffering from this form of anemia produce hemoglobin *S*. The disease takes its name from the fact that the red blood cells of persons with sickle-cell anemia become sickle-shaped under conditions of low oxygen concentration.

Genetic analysis of pedigrees from families in which certain individuals possess an abnormal form of hemoglobin has revealed a pattern of inheritance that is characteristic of the Mendelian inheritance of a single gene.

Hemoglobin *A* (and its mutant forms) is comprised of four polypeptide chains; namely, two α chains and two β chains. The properties

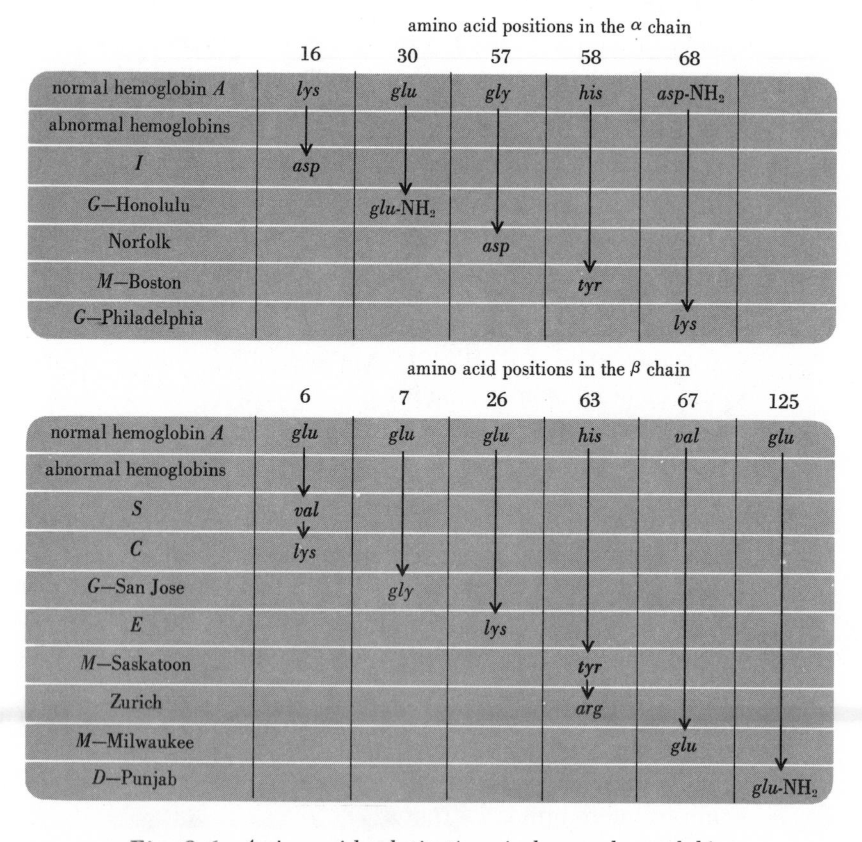

Fig. 9-6 *Amino acid substitutions in human hemoglobin.*

of the α chain are determined by the α gene and the properties of the β chain are determined by the β gene so that a mutation in the α gene leads to an aberrant α polypeptide and a mutation in the β gene leads to an aberrant β polypeptide.

Ingram undertook to determine the amino acid sequence of the α and β chains of hemoglobin. He found that the amino acid sequences of hemoglobin *A* and the two abnormal hemoglobins *S* and *C* differed within a fragment of the β chain. The difference between hemoglobin *A* and hemoglobin *S* was in the substitution of valine for glutamic acid at position 6 in the chain. The substitution in hemoglobin *C*, also at position 6, was lysine for glutamic acid. More than a dozen abnormal hemoglobins have been discovered, and all differ from hemoglobin *A* by *single amino-acid substitutions* at specific sites in either the α or β chains (Fig. 9-6).

Similar observations have been made with regard to the protein that comprises the coat of TMV. The amino acid sequence of the protein of wild-type TMV is known, and when mutants were induced the amino acid sequence of their protein was compared with the wild-type protein. Single amino-acid substitutions were found in each case, illustrating again that mutations lead to a change in protein primary structure.

A third example is that of the *A* polypeptide chain of the enzyme tryptophan synthetase of *E. coli.* C. Yanofsky and his co-workers have shown that single amino-acid substitutions have occurred in the *A* chain in several of the mutant strains that lack tryptophan synthetase activity.

A MOLECULAR INTERPRETATION OF THE MUTATION PROCESS

At the conclusion of Chapter 1 we learned that the DNA double helix was composed of a sequence of purine-pyrimidine pairs, and that during DNA replication, specificity of pairing ensured genetic continuity between cell generations. When the structure of DNA was established, it also became clear that the linear sequence of bases could comprise a genetic code, the code deriving from the many possible different sequences of bases in the molecule.

Let us assume that a given gene possesses a specific sequence of base pairs, and that this sequence contains the information required for the function of the gene. This function, in turn, plays an essential role in the development of the phenotype characteristic of the gene in question. Accordingly, an alteration in the sequence could change the information, and thus the function. Ultimately there would be a change in the phenotype. On the basis of these assumptions, mutation would be the result of a change in the sequence of purine-pyrimidine pairs within a DNA molecule.

In their hypothesis for the structure of DNA, Watson and Crick proposed that during DNA replication, mistakes in pairing might occur between certain of the bases. If a mistake in pairing were to lead to an alteration in base sequence, then on the basis of the assumptions we have made, a mutation could have occurred.

Mutations could occur by other means as well. If, for example, at a given point in the sequence of nucleotide pairs a pair were to be deleted or added, the result would be a change in the sequence. Mutation could also occur if a given base within the DNA were to undergo a chemical change so that its pairing properties were different. Another way for mutation to occur would be via the insertion into the DNA of some foreign or unnatural base whose pairing properties would be different from the native or natural base. A variety of different mutagenic agents have been used in an effort to determine the molecular basis of mutation, and the most informative experiments have been those using mutagens having an action that is known to be specific on DNA. Let us examine some of the results of these experiments and see how they are interpreted in terms of possible changes in base sequence.

Hydroxylamine The compound hydroxylamine (NH_2OH) is an effective mutagen for certain bacteriophages. It can act as an agent that causes the alteration of cytosine to form a base that pairs with adenine. Consider part of the base sequence within a given gene to be

A–T
C–G
T–A
G–C

where A = adenine, T = thymine, G = guanine, and C = cytosine. If phages are exposed to hydroxylamine a given C might be altered to form a base that will be called B. This base has pairing properties that are similar to thymine. The second nucleotide pair in the sequence given above (C–G) could therefore be changed to a B–A pair. This is shown diagrammatically in Fig. 9-7, in which the nucleotide pair in question is the one in boldface type. The cytosine of the second nucleotide pair is altered to form B. When the DNA duplicates, B will pair with adenine and, as a consequence, a segment of one daughter DNA molecule will be formed having a base sequence that differs from the parental segment. The reason is that a C–G pair has been changed to a T–A pair. This sort of change is called a *transition.*

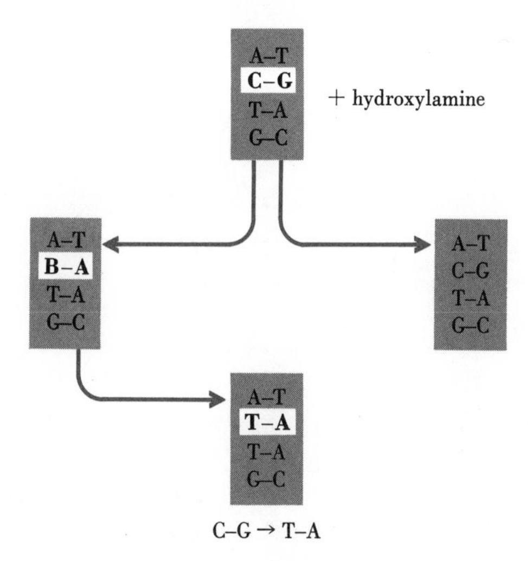

Fig. 9-7 *The changes in base sequence brought about by hydroxylamine (see text for details).*

Base Analogues Transitions are also induced by base analogues. An analogue is a compound whose molecular structure is very similar to some natural compound, and which, in a biological system, is active as an antagonist to it. In many cases the analogue can take the place of the natural substance in some important biochemical reaction. For example, the drug sulfanilamide is an analogue of the vitamin *para*-aminobenzoic acid. They are similar in structure, and in the presence of an appropriate concentration of sulfanilamide many microorganisms die because the sulfanilamide replaces the *para*-aminobenzoic acid in an essential reaction.

There are many purines and pyrimidines other than the few that are found in DNA and RNA, and a number of them occur naturally within the cell. Others, however, are unnatural analogues that have been synthesized in the laboratory.

The structure of 5-bromouracil (BU), an analogue of thymine, is shown in Fig. 9-8. This analogue is known to be mutagenic in bacteriophage and bacteria, and there is evidence to show that an analogue of adenine, 2-aminopurine, is mutagenic not only in phage and bacteria but in algae as well.

When *E. coli* is deprived of thymine and grown in the presence of BU, the BU will be incorporated into the DNA, and under appropriate conditions all of the thymine can be replaced with the analogue. If these bacteria are then infected with wild-type phage T_4, a number of phage progeny are recovered that are not wild type but *r*-type mutants.

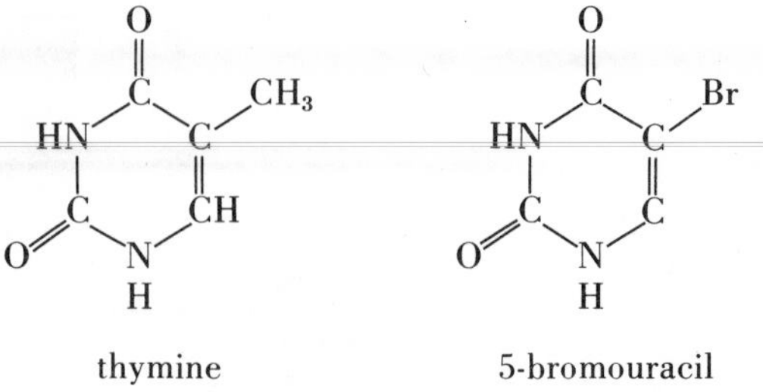

thymine

5-bromouracil

Fig. 9-8 Structural formulas of thymine and 5-bromouracil.

The frequency at which *r* mutations are induced by BU has been estimated to be several hundred times that of the spontaneous frequency. The analogue, therefore, is a very effective mutagen.

In his investigation of the induction of mutations with base analogues, E. Freese was able to take advantage of the ease with which *r* mutants can be mapped in T_4. Using the technique described in Chapter 8, Freese was able to localize a large number of spontaneous mutants, as well as those induced by BU. Figure 9-9 shows some of the points along the genetic map at which *r* mutants have been recorded as having arisen spontaneously or in the presence of BU. There are two striking things to be noted in these maps. First, there are some sites at which mutations appear to occur very frequently. Such "hot spots" are seen in each of the maps. Second, the distribution of the mutations, as well as the "hot spots, " is different in each of the maps. Thus, the mutation "spectrum" is different for the mutants that occur spontaneously in comparison with those occurring in the presence of BU. The fact that each of these maps has a different distribution of mutations, as well as characteristic "hot spots," suggests the possibility of a different mode of origin for the spontaneous mutations compared with those induced by the base analogue.

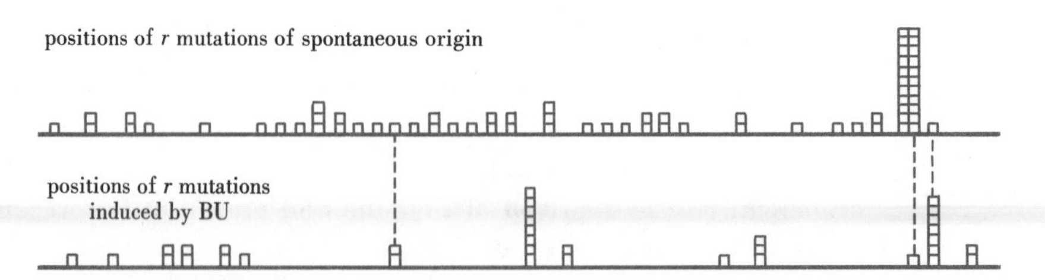

Fig. 9-9 Genetic map of spontaneous and 5-bromouracil-induced r *mutants in phage. (After Benzer and Freese,* Proceedings of the National Academy of Sciences, *Vol. 44, 1958.)*

An investigation of the reverse mutation from $r \rightarrow +$ has revealed an important difference between spontaneous and analogue-induced mutations. Both mutations can revert spontaneously at a low rate. However, mutations that have been induced by BU can be caused to revert at an increased rate by treatment with the base analogue, whereas the reversion of mutants that arose spontaneously is not appreciably affected by the base analogue.

These results have suggested a thought-provoking hypothesis for the mutation process. Let us, as before, assume that a mutation will occur when the sequence of nucleotide pairs becomes altered within the DNA representing the gene in question. Consider the sequence within this particular gene to be

A–T
G–C
A–T
C–G

A mutation could be a change in sequence—as, for example,

A–T
G–C
G–C
C–G

where the third nucleotide pair has been changed from A–T to G–C.

We must now ask how a base analogue might bring about such a change in sequence, and hence a mutation. Since it is known that an analogue such as BU can be incorporated into DNA, the specific question would seem to be whether or not its incorporation can have any effect in shifting an A–T pair to a G–C pair. The mere replacement of thymine by BU is insufficient to cause mutation, since it has been shown that DNA duplication is necessary for a mutation to occur. It has been suggested, therefore, that the mutations induced by BU are a form of copy error that occurs during DNA duplication. There are two kinds of errors: one that arises when A pairs initially with BU; and the other, when G pairs initially with BU (Fig. 9-10). The first sort of error is shown in Fig. 9-10A, in which the nucleotide pair in question is the one in boldface type. It can be seen that an A–T pair is exchanged for a G–C pair in a process that requires three steps. First, A becomes paired with BU at the first DNA replication. Second, BU then forms a pair with G at the next replication. Third, at the next replication G pairs with C. This change from A–T to G–C has been effected through the error in the second step when the analogue pairs with G instead of A.

The second kind of error, in which G pairs initially with BU, is shown in Fig. 9-10B. It is seen that the result is a change from a

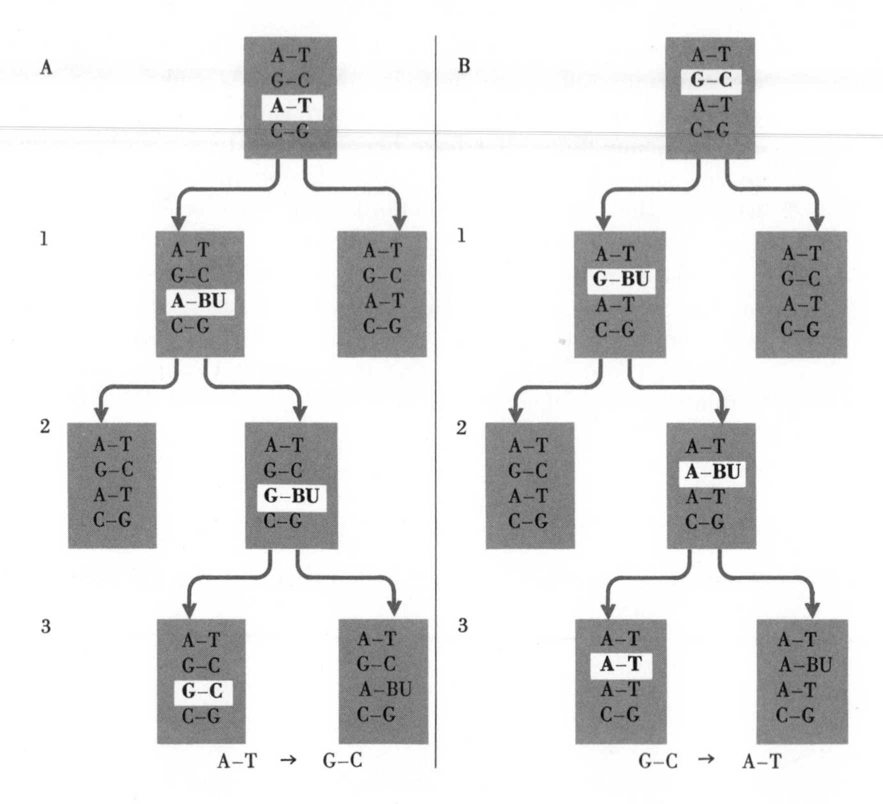

Fig. 9-10 The changes in base sequence brought about by pairing between BU and adenine and between BU and guanine (see text for details).

G–C pair to an A–T pair. Thus, according to this hypothesis, the analogue is capable of bringing about changes in two directions: A–T → G–C and G–C → A–T (or T–A → C–G and C–G → T–A). Let us assume that a forward mutation is one in which the analogue brings about the change A–T → G–C. Accordingly, the reverse mutation would be G–C → A–T. The hypothesis described here indicates how the analogue might bring about both of these mutations (*A–T ↔ G–C*) and thus it fits the experimental data obtained from the phage experiments. It was pointed out, however, that spontaneous mutations do not revert readily in the presence of base analogues. Therefore, a spontaneous mutation may represent some sort of alteration in base pairs other than A–T → G–C.

It is significant that a mutation induced by hydroxylamine can be induced to revert in the presence of a base analogue such as BU but not by hydroxylamine. This finding is predicted from the hypothesis for base analogue-induced mutations, since hydroxylamine and BU can induce C–G → T–A transitions but hydroxylamine cannot induce a T–A → C–G transition.

Ultraviolet Irradiation *In vitro* experiments with DNA have shown that UV-induced mutations can arise in several ways. Upon UV irradiation, chemical bonds can be formed between adjacent pyrimidines in one chain of the helix, for example, between cytosine and thymine (a cytosine-thymine *dimer*). The hydrated form of a pyrimidine can also be formed, for example, cytosine can be hydrated. In both cases, following dimerization or hydration, the cytosine can become converted to uracil. Thus, uracil now occupies the position instead of cytosine. Accordingly, at replication the uracil will pair with adenine and at the next replication the adenine will pair with thymine. Thus, a C–G pair becomes changed to a T–A pair.

Genetic experiments with phage have demonstrated that many UV-induced mutations are of two kinds. The first kind can be interpreted as a base-pair transition from G–C to A–T because, just as with mutations induced by hydroxylamine, these mutations are revertible with base analogues. The second class of mutations consists of either deletions or additions of from one to twenty base pairs. As expected, these mutations are not revertible with base analogues but they are revertible with the acridine dye proflavin.

Acridine Dyes The induction of mutations by causing base-pair transitions is to be contrasted with the mechanism of induction of mutations by certain acridine dyes. The acridine dye, proflavin, is an effective mutagen in phage T_4, where it has been used to induce *r* mutations. Interestingly, almost all of the *r* mutations induced by proflavin occur at sites that are distinct from those of mutations induced with other mutagens. These proflavin-induced mutations will revert spontaneously but they are not caused to revert with base analogues. Thus, the proflavin-induced mutations behave in a fashion that is similar to spontaneous mutations.

Although the mode of action of proflavin is not completely understood, it has been suggested that mutations induced by acridine dyes, as well as those of spontaneous origin, are not transitions but result instead from either the addition or the deletion of a base pair. It is known that acridines (at the right concentration) may become inserted between adjacent base pairs of the DNA double helix, making the molecule longer and stiffer. The insertion of one or more proflavin molecules can lead to misalignment of DNA molecules during the process of genetic recombination. As a result, imperfect recombination events ensue, and thus the recombinant molecules possess from one to perhaps twenty *too few* or *too many base pairs* at the region where recombination occurred. The change induced by proflavin is, therefore, a consequence of recombination; it results in either the deletion or addition of a base pair or pairs. The induction of mutations by either the addition or the deletion of base pairs has played an important role in the attempts to understand the nature of the genetic code and its mode of operation in determining the synthesis of proteins. The code and its operation are considered in the next chapter.

FURTHER READING

Hayes, W., 1964. *The genetics of bacteria and their viruses.* Oxford: Basil Blackwell & Mott, Ltd.

Swanson, C., 1957. *Cytology and cytogenetics.* Englewood Cliffs, N.J.: Prentice-Hall, Inc.

Watson, J. D., 1965. *The molecular biology of the gene.* New York: W. A. Benjamin, Inc.

chapter 10

DNA, RNA, and Protein Synthesis

The mutation of a gene can affect an organism's ability to synthesize a given protein. For example, a mutant organism may fail to synthesize an enzyme or it synthesizes a protein similar to the enzyme but modified in a way so as to be enzymatically inactive. From this observation it can be concluded that there is an intimate relationship between a given gene and a given protein. Furthermore, it is known (see Chapter 9) that the mutant form of a gene can arise because of an alteration in the nucleotide sequence of DNA. Thus, it is suggested that the *sequence of nucleotides* is important in determining the sort of protein to be synthesized. Indeed, as was pointed out in Chapter 9, it has been demonstrated in the cases of human hemoglobin, tryptophan synthetase from *E. coli*, and the coat protein of TMV that the proteins synthesized by the mutant forms of these organisms have an altered amino-acid sequence.

It is now clear that the nucleotide sequence of DNA can direct the assembly of amino acids into the specific sequence that represents the primary structure of a protein. This assembly depends on the nature and operation of the genetic code inherent in the nucleotide sequence of DNA, and on the way that the information of the code becomes translated into an amino acid sequence. The overall process of protein synthesis is viewed in terms of two major steps. The first is the *transcription* of the code from a sequence of nucleotides in one strand of DNA to the *complementary* sequence of nucleotides in RNA and, in particular, the nucleotide sequence of a species of RNA called *messenger* RNA or *m*-RNA. The second major step is the *translation* of the nucleotide sequence of *m*-RNA by a process that assembles amino acids into a specific sequence. These steps will be discussed in detail after we have examined the nature of the genetic code.

THE GENETIC CODE

Several different types of codes have been proposed to explain how sequences of nucleotides might be used in coding the amino acids into a protein. Twenty different amino acids are commonly found in proteins, and four different bases are found in DNA and RNA. One problem, therefore, is how many bases are necessary to code for each of the twenty different amino acids. In other words, what grouping of nucleotides comprises a code word or *codon*? Of the several codes that have been proposed since the establishment of the Watson-Crick structure for DNA, only the *nonoverlapping triplet code* has found support from experimental evidence of both a genetic and biochemical sort. This is a code in which a codon is composed of three nucleotides or a *triplet*. With four bases, 4^3 or 64 different triplets are possible. Since there are only twenty different amino acids, twenty triplets would, therefore, seem sufficient to code the different amino acids. The remaining 44 triplets could be unnecessary or serve some other function. On the other hand, more than one kind of triplet could code for the same amino acid and, therefore, the code would be *degenerate*.

Evidence for the nonoverlapping nature of the code comes from the analysis of the amino-acid sequence of mutant proteins such as the coat protein of TMV and the tryptophan synthetase of *E. coli*. For example, the mutations of TMV that alter the amino acid sequence of the coat protein are, for the most part, changes in only a single amino acid. When, in very rare instances, two amino acids are changed, they are not in adjacent positions. If the code were an overlapping triplet code, a change in one of the nucleotides of a codon could change as many as three adjacent amino acids in the protein (Fig. 10-1). Similar observations have been made with respect to *E. coli* tryptophan synthetase.

Evidence that triplets of nucleotides could account for the coding of a single amino acid came first from genetic experiments by Crick and his co-workers using *r* mutants of phage T_4. Proflavin was used to induce mutations in the *r*II gene, and the mutants were mapped according to the procedure described in Chapter 8. Mutations mapping in different parts of the *r*IIB cistron of the *r*II gene were selected for investigation.

It will be recalled from Chapter 9 that proflavin can cause either the insertion or the deletion of nucleotide pairs. Further, these mutations can revert either spontaneously or in the presence of proflavin. Among the proflavin-induced mutations in the *r*IIB cistron, several spontaneous and proflavin-induced wild-type revertants were obtained. However, when they were crossed with wild type it was found that they were not true revertants but double mutants; a second mutation had occurred at a different site in the *r*IIB cistron which was mutant by itself, but when it was combined with the original mutation, a wild phenotype resulted. Thus, in the double mutant some sort of correction process occurred to produce the wild phenotype. These mutations that restore the wild phenotype when in combination with the original mutations are called *intragenic suppressors*, and they can be denoted by the symbol +. The original mutations are given the symbol −. By appropriate crosses various combinations of the *r*IIB mutations were made, and it was established that each single + or − mutation, and each double + or double − mutation gave mutant phenotypes. Four or five − mutations also gave a mutant phenotype. However, when − + combinations were made, the phenotype was, or approached that of, wild type. This was also true for most triple − combinations, as well as for multiples of three − mutations, and for combinations of three + mutations.

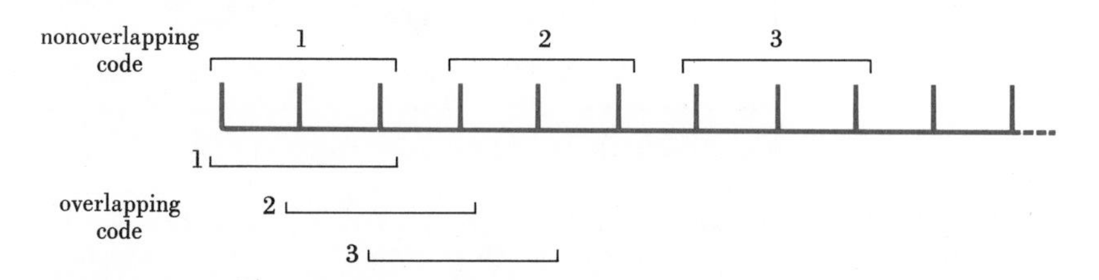

Fig. 10-1 *Overlapping and nonoverlapping genetic codes. (After Crick, Brenner, and Watts-Tobin,* Nature, *Vol. 192, 1961.)*

The interpretation of these results can be best understood by following the diagrams in Fig. 10-2, in which it is assumed that a − mutation represents a proflavin-induced deletion of a nucleotide pair and a + mutation represents a proflavin-induced insertion of a nucleotide pair. In these diagrams only part of *one* strand of the DNA molecule is shown; it is known that only one strand of the DNA molecule serves as the template for RNA synthesis.

Figure 10-2A represents a portion of the phage T_4 DNA that can be assumed to code for part of the protein synthesized under control of the *r*IIB cistron. As shown, the code reads from left to right, CAT, CAT, and so on. Figure 10-2B shows what occurs if a nucleotide is *deleted.* The code reads CAT, CAT, and so on, for the first four codons. However, as a result of the deletion at the fifth codon, the reading shifts to ATC, ATC, and so on, beginning with the sixth codon. The protein thus synthesized would contain, from this point on, an amino-acid sequence different from that of the wild-type protein. Such a sequence is either a *missense* sequence if ATC codons code for a different amino acid, or a *nonsense* sequence if ATC codes for no amino acid. Similarly, as shown in Fig. 10-2C, the *insertion* of a nucleotide in the third codon would change the reading at the fourth codon and onwards. If, however, both a deletion and an insertion were to occur, as in the − + double

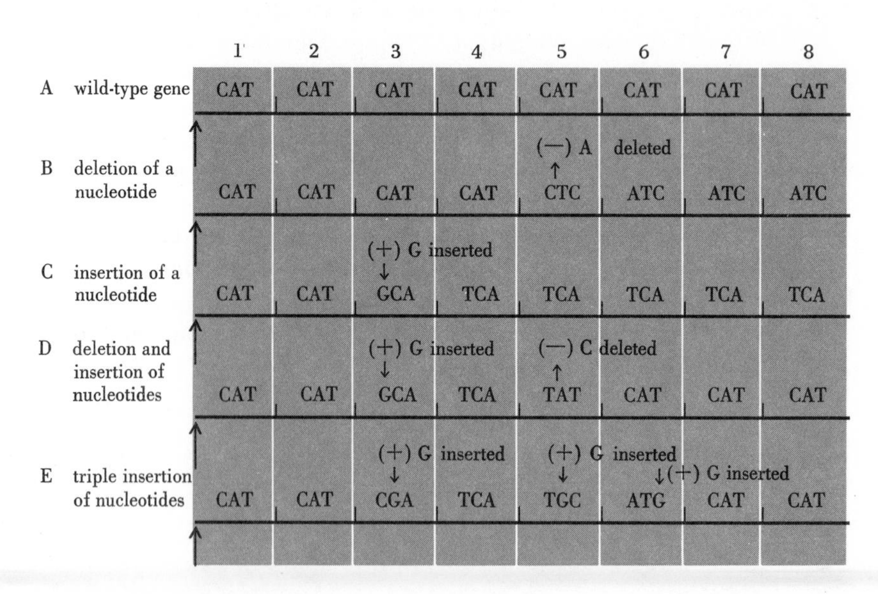

Fig. 10-2 *The reading of the genetic code. The nucleotide sequence CAT, CAT, and so forth, is read from left to right. The wild-type sequence (A) has been altered by the deletion or insertion of nucleotides and consequently the reading of the code will be changed. Further details in the text. (After Crick, Brenner, and Watts-Tobin,* Nature, *Vol. 192, 1961.)*

mutant shown in Fig. 10-2D, the reading of the code would be restored to the proper order beginning with the sixth codon. Now, assuming that the aberrant order were to code for a portion of the *r*II protein in which missense was not deleterious—that is, had no major effect upon the activity of the protein—the correct reading after the deletion could lead to the synthesis of a protein having at least partial wild-type activity. Clearly, the restoration of the wild-type activity depends upon the positions of the mutations with respect to each other. If, for example, the insertion and deletion were relatively far apart, the probability of obtaining a wild-type protein would decrease because the number of altered codons had increased.

The observation that three − or three + mutations can restore the function of the *r*IIB cistron but that two − or two + mutations will not implies that the third mutation restores the reading of the code. This is shown in Fig. 10-2E. The same result would be obtained if three − mutations were included in the *r*IIB cistron. The finding that triple + or − mutants have the wild phenotype provides strong support for the triplet nature of the code. However, it does not necessarily exclude the possibility that the codons are some multiple of three.

Given a triplet code, one of the important questions is the assignment of codons which specify the incorporation of different amino acids into a protein. In order to answer this question it is necessary to have a knowledge of the way in which proteins are synthesized.

PROTEIN SYNTHESIS

Protein synthesis can be obtained *in vitro*, and there is little reason to doubt that the *in vitro* synthesis exemplifies *in vivo* protein synthesis. The *in vitro* synthesis of proteins has several requirements. Amino acids must be provided, as well as ATP as an energy source. In addition, ribonucleoprotein particles called *ribosomes*, a low molecular weight RNA called *soluble* RNA, and an RNA called *messenger* RNA are required. These may be isolated from cells of a variety of organisms. Certain enzymes are also required for protein synthesis. The entire process is ultimately dependent upon DNA.

DNA has two functions. First, it serves as a template for its own duplication, and second, it serves as a template for the synthesis of RNA. In its second function the genetic information for protein synthesis, inherent in the nucleotide sequence of DNA, is *transcribed* into the nucleotide sequence of messenger RNA. Only one strand of the DNA molecule serves as the template in this transcription process. The protein-synthesizing system then *translates* the messenger RNA nucleotide sequences into the amino-acid sequence of a protein.

The three different types of RNA molecules play specific roles in protein synthesis. The bulk of the RNA is called *ribosomal* RNA (*r*-RNA), and it is found in the ribosomes. Gentle disruption of cells reveals that the ribosomes can be found in aggregates called *polyribosomes* or *polysomes* (see Fig. 10-3), and it is known that the assembly of amino acids into proteins occurs on the surface of the ribosomes.

The second type of RNA is a relatively small molecule called *soluble* RNA or *s*-RNA (in some instances it is referred to as *transfer* RNA or *t*-RNA). The *s*-RNA takes part in a step in protein synthesis known as *amino acid activation.* In this step an individual amino acid becomes covalently bound to one end of an *s*-RNA molecule by an enzymatic reaction. The activation of a given amino acid involves an amino-acid-activating enzyme and an *s*-RNA, both of which are specific for that amino acid. There are twenty specific amino-acid-activating enzymes, and at least twenty specific types of *s*-RNA molecules, each of which has the nucleotide sequence of cytosine, cytosine, adenine (CCA) at the 3′ terminal end and guanine (G) at the 5′ terminal end. An amino acid becomes covalently bound to its *s*-RNA at the 3′ end that terminates in adenine. The specifity of this process is mediated by the amino-acid-activating enzyme so that, for instance, only tyrosine *s*-RNA is able to find an activated tyrosine molecule.

The next step in protein synthesis is the one in which the amino acids become joined together in a specific sequence. This step requires *m*-RNA. Many different *m*-RNA molecules are produced by a cell, presumably at least one for each kind of protein to be synthesized. Each *m*-RNA molecule has a nucleotide composition and sequence that is *complementary to that of the DNA template on which it was formed.* The nucleotide sequence of an *m*-RNA molecule endows it with the information necessary for ordering amino acids into their proper sequence, and specifically, the information lies in *triplets* of nucleotides each representing a code word corresponding to a specific amino acid. Each *s*-RNA, in turn, carries a triplet of nucleotides, termed the *anticodon*, which is complementary to the *m*-RNA codon for tyrosine. This *s*-RNA-*m*-RNA pairing effects the ordering of amino acids in the protein. Apparently the *s*-RNA anticodon "finds" the appropriate *m*-RNA codon by a random process.

Protein synthesis entails the linear growth of a chain of amino acids called a *polypeptide.* One end of the chain is called the *C* terminal end, that is, an end terminating in a carboxyl group. The other end is an *N* terminal end, which terminates in an amino group. It has been found that during protein synthesis the growth of a polypeptide chain commences with its *N* terminal end.

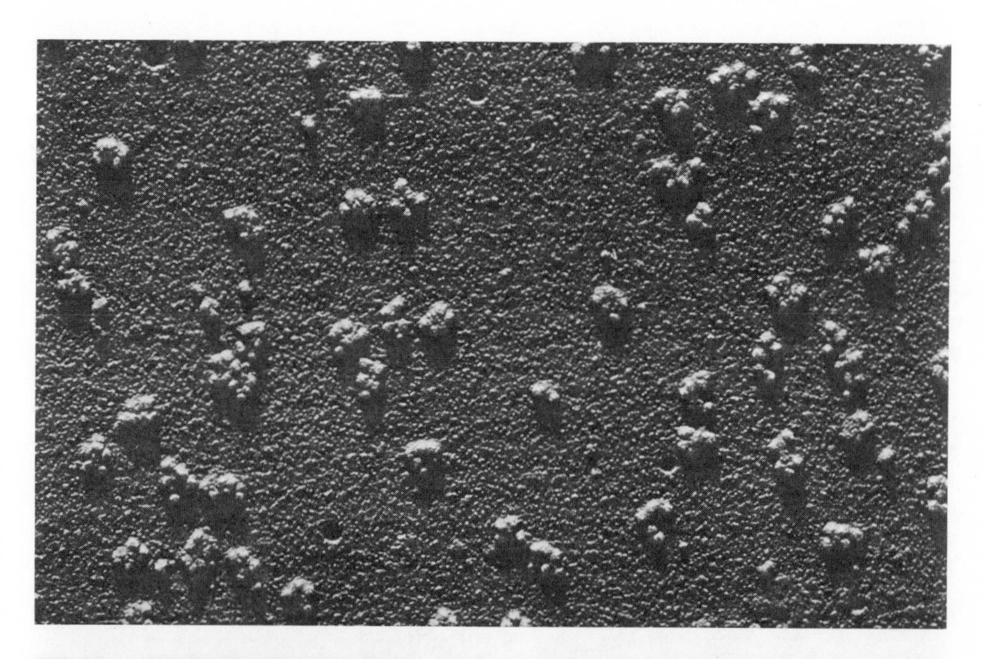

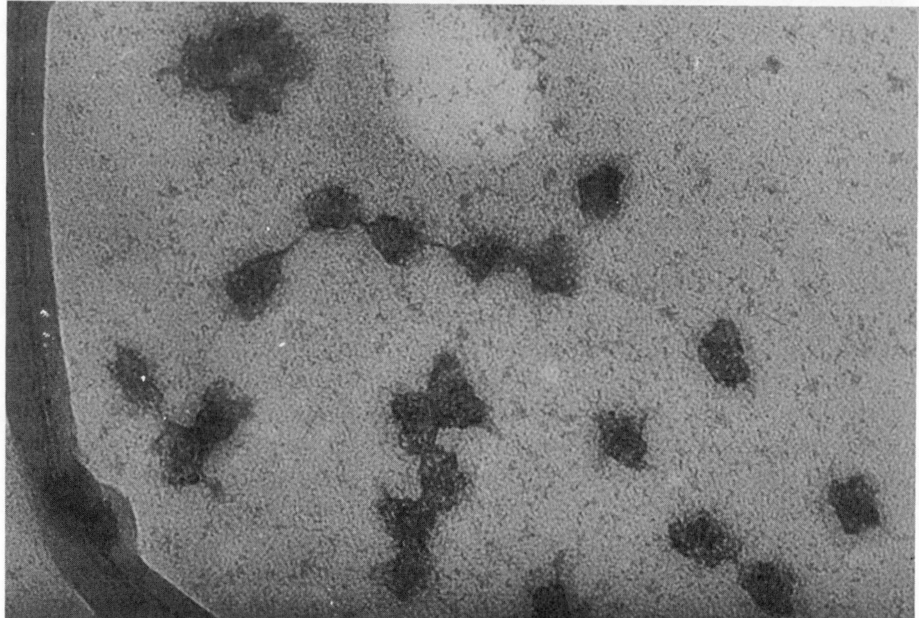

Fig. 10-3 *Polyribosomes from rabbit reticulocytes. Clusters of four to six ribosomes are shown in the top electron micrograph. The lower electron micrograph reveals that there are threads between some of the ribosomes. These threads may be* m-*RNA. (Photographs courtesy of A. Rich.)*

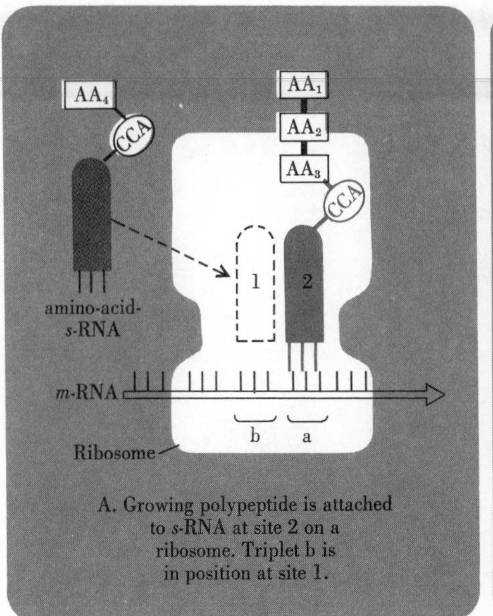

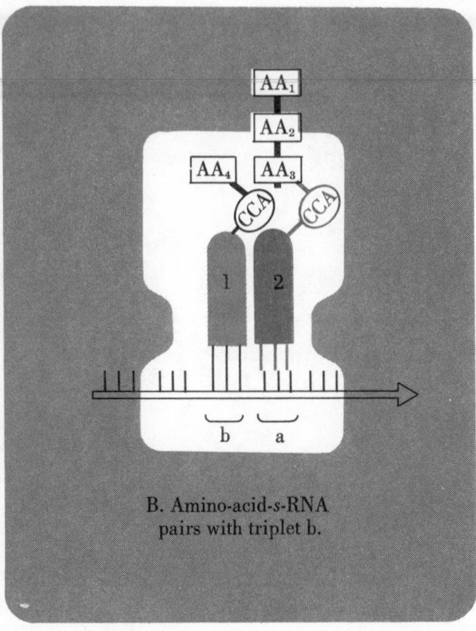

Fig. 10-4 *Steps in polypeptide synthesis. (After Watson,* Molecular Biology of the Gene. *New York: W. A. Benjamin, Inc., 1965).*

The assembly of a polypeptide chain is diagrammed in Fig. 10-4. The ribosome is thought to have two *s*-RNA-specific sites. At one site, amino-acid-*s*-RNA molecules become temporarily bound to their complementary *m*-RNA triplets. At the second, the growing peptide chain is attached to a second *s*-RNA molecule (Fig. 10-4A). The *m*-RNA moves along the surface of the ribosome, each of its triplets being momentarily positioned at the first site. When an appropriate amino-acid-*s*-RNA arrives at this site, it pairs with its complementary *m*-RNA triplet (Fig. 10-4B). A peptide bond is then formed by an enzymatic process, adding a new amino acid to the growing peptide chain. As this occurs, the original *s*-RNA molecule loses its connection with the peptide chain and leaves the ribosome (Fig. 10-4C); it is now free to bind with another amino acid. In the final step, the *m*-RNA and its associated peptide chain moves along the ribosome, exposing a new triplet at the first site (Fig. 10-4D).

In the example, protein synthesis was described as occurring on only one ribosome. In fact, *m*-RNA can become associated with a group of ribosomes, forming the polyribosome described earlier (Fig. 10-3). The *m*-RNA moves along this group of ribosomes, and synthesis of several polypeptide chains can occur simultaneously.

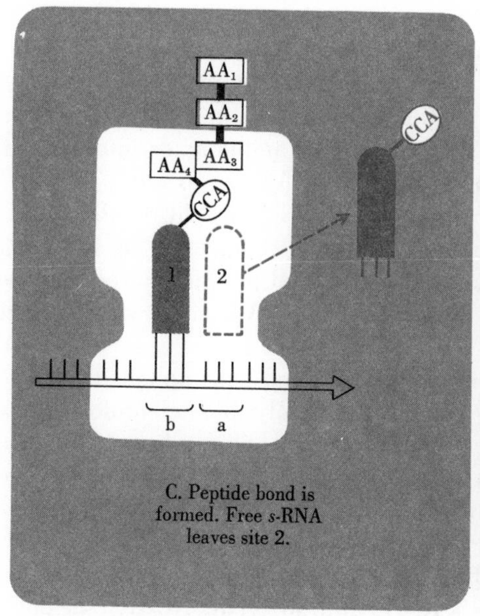

C. Peptide bond is formed. Free *s*-RNA leaves site 2.

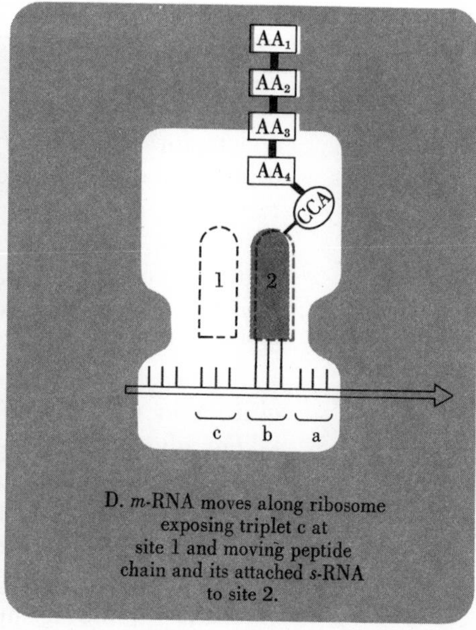

D. *m*-RNA moves along ribosome exposing triplet c at site 1 and moving peptide chain and its attached *s*-RNA to site 2.

THE ASSIGNMENT OF CODONS

When *in vitro* protein synthesis is studied using natural *m*-RNA, a variety of different amino acids become incorporated into a protein. However, it is possible to produce synthetic polyribonucleotides that can function as *m*-RNA. A significant feature of the synthetic polyribonucleotides is that they can be made with a known base composition. For example, it is possible to make polyuridylic acid (poly U), polyadenylic acid (poly A), and polycytidylic acid (poly C). With poly U as the *m*-RNA in a protein synthesizing system, the polypeptide formed contains only one type of amino acid, namely phenylalanine. This result suggests that uracil triplets can code for phenylalanine. With poly A as *m*-RNA a polypeptide is formed containing only the amino acid lysine, and when poly C is used the polypeptide is that of proline. Therefore, it appears that adenine triplets can code for lysine, and cytosine triplets can code for proline.

What happens when mixed polymers are made? It is possible to synthesize polyadenine-cytosine (poly A–C); when this is used as the *m*-RNA, six different amino acids are incorporated into the polypeptide: asparagine, glutamine, histidine, lysine, threonine, and proline. The next step is to synthesize mixed polymers with different ratios of A and C, and it is found that when there is more A than C the ratio of asparagine to histidine in the polypeptide increases. Finally, a mixed polymer can be synthesized in which there is a regular base sequence, for example, a regular copolymer of C and U (CUCUCUCU...). When the C–U copolymer is used as the *m*-RNA, two different amino acids, leucine and serine, are incorporated into the polypeptide.

Let us now turn to the question of the assignment of *specific* codons for the incorporation of specific amino acids into protein. Keep in mind that a given codon in DNA (CAT, for example) is transcribed to *m*-RNA as the complementary GUA. You will recall that G and C are complementary bases, and that U replaces T in RNA. Thus, A and U are complementary. It is the usual practice to refer to codons in terms of the RNA bases instead of the DNA bases of which the ultimate codons are comprised.

The assignment of codons has been accomplished by the use of *m*-RNA in the form of the synthetic polyribonucleotides and also by taking advantage of the fact that *s*-RNA will bind to ribosomes in the presence of a trinucleotide in a system lacking the other requirements for protein synthesis. It is possible to isolate, purify, and to identify and separate each different *s*-RNA on the basis of the specific amino acid that it binds. In other words, it is possible to obtain the phenylalanine *s*-RNA, the serine *s*-RNA, and so forth. Next, it is possible to

Table 10-1 **The Assignment of Codons**

First Position (5′ End) (Read Down)	*Second Position (Read Across)*				*Third Position (3′ End) (Read Down)*
	U	C	A	G	
U	*phe*	*ser*	*tyr*	*cys*	U
	phe	*ser*	*tyr*	*cys*	C
	leu	*ser*	Stop	*tryp*	A
	leu	*ser*	Stop	*tryp*	G
C	*leu*	*pro*	*his*	*arg*	U
	leu	*pro*	*his*	*arg*	C
	leu	*pro*	*glun*	*arg*	A
	leu	*pro*	*glun*	*arg*	G
A	*ileu*	*thr*	*aspn*	*ser*	U
	ileu	*thr*	*aspn*	*ser*	C
	ileu	*thr*	*lys*	*arg*	A
	meth (start)	*thr*	*lys*	*arg*	G
G	*val*	*ala*	*asp*	*gly*	U
	val	*ala*	*asp*	*gly*	C
	val	*ala*	*glu*	*gly*	A
	val	*ala*	*glu*	*gly*	G

synthesize a variety of different trinucleotides. For example, one can make the CCC and UUU trinucleotides as well as those that have all possible sequences of C and U (that is, CCU, CUC, CUU, UUC, UCU, UCC). Each trinucleotide and each *s*-RNA can then be tested in the presence of ribosomes to detérmine which trinucleotide is responsible for the binding of which *s*-RNA to the ribosomes. The assignment of codons by this method, as well as by the use of synthetic polynucleotides, gives the assignments shown in Table 10-1 for the twenty different amino acids. As you will notice, the assignments verify the idea that the code is degenerate. You will also notice that two codons, UAA and UAG, have not been assigned to a given amino acid. These codons serve as the stop signal for the formation of a polypeptide chain. Later on in this chapter we shall learn more about the function of these codons.

Although it has been possible to assign codons for the twenty amino acids the code is not unambiguous, for certain codons can code for the wrong amino acid. For example, the codon for phenylalanine (UUU) can, under the appropriate experimental conditions, code for isoleucine. There are other ambiguities as well and therefore the code is not perfect.

THE COLINEARITY AND UNIVERSALITY OF THE CODE

Once we have established that the code is a triplet code, it is essential to determine whether or not there is *colinearity* between a gene and its protein product. In other words, both the gene and its product are linear structures, and a mutational change at a given site within the gene should lead to an amino acid substitution at a position in the protein that is the same position relative to the mutational change (Fig. 10-5).

In Chapter 9 it was pointed out that mutation has been shown to lead to amino-acid substitutions in three proteins: human hemoglobin, tryptophan synthetase of *E. coli*, and the coat protein of TMV. The extensive genetic and biochemical investigation of mutants of *E. coli* that lack tryptophan-synthetase activity has revealed that there is a colinear relationship between the gene and its product, tryptophan synthetase.

Tryptophan synthetase of *E. coli* can be separated into two polypeptide chains (chains A and B), and the amino acid sequence of each chain is determined by two adjacent genes, the A and B genes. A large portion of the amino acid sequence of the A chain of the wild-type tryptophan synthetase is known. Among the many mutant strains that lack tryptophan-synthetase activity there are several that synthesize

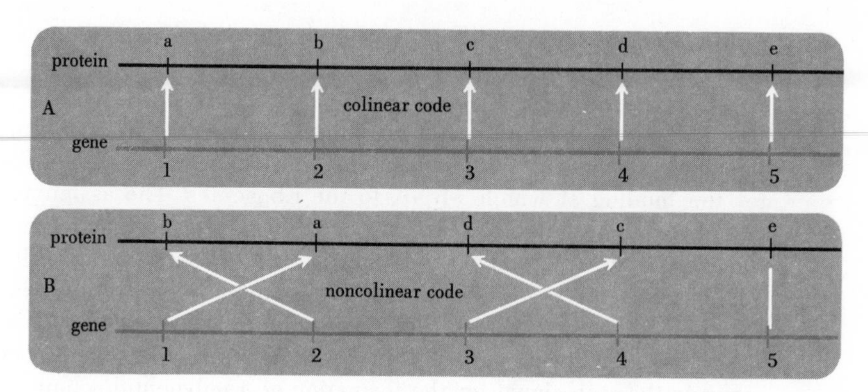

Fig. 10-5 *In a colinear code the position of a given amino acid* (a–e) *in a protein corresponds to the position of a given nucleotide (1–5) in a gene (all arrows vertical in part A). In a noncolinear code the position of a given amino acid in a protein* does not *necessarily correspond to the position of a given nucleotide in a gene (some arrows are diagonal in part B and they cross).*

an inactive A chain, and each of these differs from the wild type by a single amino-acid substitution. The amino-acid sequence of the different mutant A chains has been determined and, significantly, the position of a given substitution corresponds exactly to the position in the genetic map at which a given mutation has occurred. Thus, two linear structures, the A gene and the A polypeptide chain are *colinear* with respect to the site of mutation and the site of amino-acid substitution.

In most organisms the universality of their biochemical processes transcends their taxonomic differences. The same appears to be true for the genetic code. For example, hemoglobin is synthesized in reticulocytes, but rabbit hemoglobin can be synthesized in an *in vitro* system containing ribosomes and *m*-RNA from rabbit reticulocytes plus *E. coli* amino-acid-*s*-RNA molecules. In addition, synthetic *m*-RNA, for example poly U, acts to code for phenylalanine in a variety of *in vitro* protein-synthesizing systems.

STARTING AND STOPPING PROTEIN SYNTHESIS

It is now possible to conceive of the gene as being that portion of a DNA molecule whose nucleotide sequence represents a code that specifies the amino-acid sequence of a specific protein. The code is *transcribed* into *m*-RNA having a complementary nucleotide sequence. The *m*-RNA then acts in the step of the *translation* of the code into specific amino acid sequences. Clearly, for any gene, both the transcription of RNA on DNA and the translation of RNA into protein must have a starting and a stopping signal. Each gene must have specific triplets that act as either starting or stopping points for *m*-RNA synthesis. There is evidence that UAA and UAG are the polypeptide-chain-terminating codons. A group of mutations has been found that gives some insight into the nature of the stop signal, at least at the level of translation.

Certain phages carry mutations that cause the synthesis of a given protein to be terminated at a certain point. It was found that when these mutant phages are grown on certain strains of *E. coli*, the chain-terminating mutations are *suppressed* and phages are produced that resemble wild-type phages.

This phenomenon has been explained by assuming that as a result of mutation, a chain-terminating codon is present in the DNA of the mutant phage at a point where an amino-acid codon occurs in the wild type. The *E. coli* strains that are able to suppress this mutation are also mutant, for they synthesize an *s*-RNA that inserts an amino acid at the chain-terminating-codon site. Consequently, synthesis of a complete protein can occur when the mutant phages are grown on the mutant hosts. This protein may resemble wild-type protein if the amino acid that is substituted for the original does not greatly alter the properties of the molecule.

The codon AUG appears to signal the start of synthesis of a polypeptide chain, and it codes for the amino acid methionine. There are two types of *s*-RNA that pertain to methionine. The function of one of them is in the insertion of methionine within a polypeptide chain. The other, following amino acid activation, also yields methionine-*s*-RNA, but this amino acid-*s*-RNA becomes formylated. That is, it carries a formyl group

$$\left(\mathrm{H}-\overset{\overset{\displaystyle \mathrm{O}}{\|}}{\mathrm{C}}-\right)$$

Polypeptide chains in an *in vitro* protein-synthesizing system have been found to begin with the formyl methionine. In about 40 percent of the cases the formyl group is removed enzymatically, and methionine is the first amino acid of the protein. In the remaining 60 percent of the cases the formyl methionine is removed enzymatically, and the protein begins with any one of several other amino acids. In the translation process, the formyl methionine-*s*-RNA binds to the AUG codon; therefore it may be that this codon can represent the start signal for polypeptide-chain formation.

Both polypeptide-chain-starting and -terminating codons become of great importance in considering those genetic systems in which the synthesis of more than one protein is coordinated by a series of linked genes. For example, in the bacterium *Salmonella* the biosynthesis of the amino acid histidine requires the participation of nine different enzymes. The genes determining the amino-acid sequence of these enzymes are linked in one very small segment of the genetic map and the sequence of the genes corresponds for the most part with the sequence in which the enzymes act. This entire gene complex transcribes a single *m*-RNA that translates the genetic information into nine different proteins. Clearly, with a single *m*-RNA molecule and with nine different enzymes to be synthesized, many chain-starting and chain-terminating codons are required in order to have the coordinated synthesis of all nine proteins. The control and coordination of enzyme synthesis are discussed in the next chapter.

FURTHER READING

Bautz, E. K. F., 1967. "RNA synthesis and the mechanism of genetic transcription," in *Molecular genetics*, part 2, p. 213. Taylor, J. H. (ed.). New York: Academic Press Inc.

Lengyell, P., 1967. "Peptide chain initiation," in *Molecular genetics*, part 2, p. 194, Taylor, J. H. (ed.). New York: Academic Press Inc.

Speyer, J. F., 1967. "Genetic code," in *Molecular genetics*, part 2, p. 137, Taylor, J. H. (ed.). New York: Academic Press Inc.

chapter **11**

Genetic Regulation

An organism is, to a large extent, a reflection of its enzymatically controlled biochemical processes. These processes, in turn, depend upon the organism's genotype and the environment in which the genotype acts. If the activity or synthesis of enzymes can be governed by genetic as well as other means, such as circumstances of the environment, then we may well conceive of growth and differentiation in terms of a chain of events in which some enzymes function at all times throughout the life of an organism, others are switched on or off as either the cellular or extracellular environment changes, and still others function only at specific places and times.

An understanding of the complex but highly coordinated processes of differentiation, at both a unicellular and a multicellular level, comes from a variety of experimental approaches. One approach is via an understanding of *genetic regulation*, a phenomenon in which certain genes regulate an organism's ability to *synthesize* a given enzyme. Important distinctions will be made between these genes and the genes with which you are more familiar—that is, those that act to determine protein primary structure.

Toward the end of the nineteenth century, biologists began to make a series of observations with microorganisms that set the stage for the discovery of genetic regulatory mechanisms. In 1899 the French microbiologist E. Duclaux observed that *Aspergillus*, growing on a medium containing sucrose as the carbon source, produced the enzyme invertase. The presence of the enzyme was not surprising, for it is required for the utilization of sucrose by catalyzing the breakdown of sucrose to glucose and fructose. However, the enzyme was not produced when the fungus was grown with either glucose or fructose as the carbon source. In other words, it was discovered that the organism produced a specific enzyme (invertase) only in the presence of a specific substrate (sucrose). This phenomenon was called *enzymatic adaptation.* It is now known that many microorganisms, and even cells of higher organisms, produce certain enzymes only in the presence of their substrates.

Not all enzymes exhibit this property. Many are produced and are active regardless of the presence or absence of substrate. For example, *E. coli* always possesses an enzyme for the breakdown of glucose, but it does not form the enzyme for the breakdown of lactose unless lactose (or certain other β-galactosides) is present in the medium. Enzymes that are always present are called *constitutive* enzymes, whereas those that are present only when their substrate is present are called *induced* enzymes.

ENZYME INDUCTION

A great deal of what we know about induced enzymes stems from the work of the French investigator J. Monod and his co-workers. In particular, much of this knowledge has been derived from an investigation of the ability of *E. coli* to use lactose as a carbon source. The utilization of the β-galactoside, lactose, requires the enzyme *β-galactosidase*, which catalyzes the breakdown of lactose to galactose and glucose. When cells of wild-type *E. coli* are grown on glucose as the carbon source, only traces of β-galactosidase activity can be detected. However, activity of the enzyme can increase by more than 1000 times when lactose is added to the medium. The presence of lactose could cause either the *activation* of existing β-galactosidase or its *de novo* synthesis.

Experiments have shown that the synthesis of the enzyme is *induced*; that is, there is *de novo* synthesis of new enzyme protein, as shown by the use of radioactive isotopes. The amino acids of the proteins of cells growing on glucose can be labeled with radioactive isotopes. These cells can then be transferred to a medium without the isotopes and with *lactose* rather than glucose as the carbon source. If β-galactosidase synthesis is induced in the presence of lactose, rather than merely activated by lactose, the β-galactosidase isolated from the cells will not be radioactive. Since there is essentially no radioactivity in the β-galactosidase, it can be concluded that enzyme induction is the synthesis of new protein. This new protein is synthesized within three minutes after the inducer is added to the medium and its synthesis ceases within three minutes after the inducer is removed.

THE GENETICS OF REGULATION

Let us now turn to the genetic evidence that relates to the induction of β-galactosidase. The utilization of lactose is dependent upon five different genes. Three of these genes are called *structural* genes because they determine the structure (that is, the amino-acid sequence) of three different proteins. One of these genes, z^+, determines the structure of β-galactosidase. The mutation $z^+ \rightarrow z^-$ leads to a loss of capacity for making an active enzyme even in the presence of the inducer (a β-galactoside); therefore, z^- cells fail to grow on lactose. The second gene, y^+, is responsible for determining the structure of an enzyme required for the transport of lactose into the cells. This enzyme belongs to a group known as *permeases*, and in this instance it is called the β-galactoside permease. The mutation $y^+ \rightarrow y^-$ results in the formation of an inactive permease; accordingly, y^- cells cannot grow on lactose. Like β-galactosidase, the synthesis of the permease is induced in the presence of lactose and certain other β-galactosides. The third gene, a, determines the structure of the enzyme transacetylase which is involved in lactose utilization. However, this gene has not been studied in detail, and it will not be included in the following discussion.

A fourth gene involved in lactose utilization is given the notation i^+. Cells that are i^+ will form β-galactosidase and the permease only in the presence of an inducer. The inducer can be, as before, the substrate lactose or other β-galactosides, some of which are not substrates for β-galactosidase. The mutation $i^+ \rightarrow i^-$ leads to a very interesting alteration in the cell's behavior, for now both the β-galactosidase and the permease are present regardless of the presence or absence of an inducer. Thus, in i^- cells both enzymes act as if they were constitutive rather than induced. The i^+ gene is called a *regulatory* gene because its function is to regulate the activity of the z^+ and y^+ genes. In cells of the genotype $i^+ z^+ y^+$, the z^+ and y^+ genes are active (that is, synthesis of β-galactosidase and permease occurs) only in the presence of an inducer, whereas in cells of the genotype $i^- z^+ y^+$, the z^+ and y^+ genes are active regardless of the presence or absence of inducer.

A fifth gene involved in the utilization of lactose, o^+, can also be considered as a regulatory gene. When o^+ mutates to o^c, enzyme synthesis is constitutive rather than induced in $i^+ z^+ y^+$ cells. The o gene is called the *operator* gene, and the meaning of this term will be brought out shortly.

From what has been described, cells that are $z^+ y^+$ can utilize lactose as a carbon source, whereas cells that are either $z^- y^+$ or $z^+ y^-$ cannot. The z and y loci are very closely linked; in fact, they are contiguous. However, they represent different cistrons as revealed by tests for complementation. Partial diploids can be formed in *E. coli* in which an F^- recipient receives a segment of DNA containing the F^+ sex factor and any one of several different genes from an F^+ donor. The segment, denoted F-m, where m stands for a given gene from the donor, may not become integrated into the recipient's genome but may reproduce autonomously in the recipient's cytoplasm. If the recipient cell is z^+ and the donor segment is F-z^-, the recipient can be considered diploid and heterozygous for the z locus; it is denoted as $\frac{z^+}{F\ z^-}$. The $\frac{z^+\ y^+}{F\ z^-\ y^-}$ diploid can utilize lactose. When the diploids $\frac{z^+\ y^-}{F\ z^-\ y^+}$ and $\frac{z^-\ y^+}{F\ z^+\ y^-}$ are formed they also can use lactose as a carbon source. Accordingly, there is complementation between z and y mutants and they are in different cistrons.

If diploids are considered with respect to the i as well as the z and y loci it is seen that i^+ is a dominant gene, since the diploids $\frac{i^-\ z^+\ y^+}{F\ i^+\ z^-\ y^-}$ and $\frac{i^-\ z^-\ y^-}{F\ i^+\ z^+\ y^+}$ require inducer for the synthesis of *both* β-galactosidase and the permease. Not only is the i^+ dominant to i^-, but i^+ must regulate activity at z^+ and y^+ by means of a product that is first in the cytoplasm and then reaches the z^+ and y^+ genes. Recall that the F segment is not a part of the recipient's genome but that it is reproduced autonomously in the recipient's cytoplasm. The i^+ product in the first diploid acts upon the z^+ and y^+ genes that are part of the recipient's genome. This product is a repressor substance. It is a protein, and it is believed to prevent the synthesis of the *m*-RNA required for the synthesis of the β-galactosidase and the permease when an inducer is absent. When an inducer is present it combines with the product of the i^+ gene to form a complex that is inactive in stopping the synthesis of the *m*-RNA for β-galactosidase and permease.

The i^+ gene regulates the activity of the z^+ and y^+ genes, but it does not do so directly; rather it acts upon the class of regulatory genes called *operator* genes. In the *lac* system of genes, the operator gene o^+ is contiguous to the *z* locus. The genetic map of the *lac* region of the *E. coli* chromosome is shown in Fig. 11-1. Whereas i^+ has been shown to be dominant and to exhibit its effect in either the *cis* or *trans* position, o^+ is capable of exerting its effect in the *cis* position only. Cells of the genotype $i^+\ o^+\ z^+\ y^+$ require an inducer for both β-galactosidase and permease synthesis, but cells of the genotype $i^+\ o^c\ z^+\ y^+$, where o^c is operator constitutive, do not. If a diploid is formed having the genotype $\frac{i^+\ o^+\ z^-\ y^+}{F\ i^+\ o^c\ z^+\ y^+}$, inducer is not required for either β-galactosidase or permease synthesis. On the other hand, the diploid $\frac{i^+\ o^+\ z^-\ y^+}{F\ i^+\ o^c\ z^+\ y^-}$ does not require inducer for β-galactosidase synthesis, but inducer is required for permease synthesis. The opposite holds true for the diploid $\frac{i^+\ o^+\ z^+\ y^-}{F\ i^+\ o^c\ z^-\ y^+}$, in that inducer is required for β-galactosidase synthesis but not for permease synthesis. Clearly, the o^+ gene has its regulatory effect only on the structural genes *in the same chromosome.*

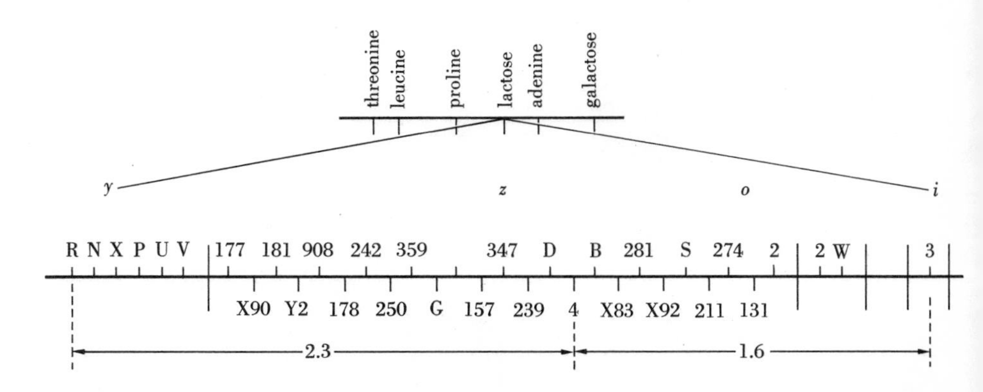

Fig. 11-1 Map of the lac *region of the* E. coli *chromosomes showing mutation in the* y, z, o, *and* i *genes.*

The i^+ product, often symbolized as R, can be thought of as a signal sent to the o^+ gene. When the signal is received by o^+, the result is that of turning "off" the z^+ and y^+ genes. If the signal is not sent, as in the case of the i^- mutation, the z^+ and y^+ genes remain "on." If an inducer is present, the receipt of the signal by o^+ is interrupted and the z^+ and y^+ genes remain "on." In the case of the o^c mutation, the signal is received but not recognized and thus the z^+ and y^+ genes remain "on."

The regulation of lactose utilization by *E. coli* is summarized diagrammatically in Fig. 11-2. The enzymes required for lactose utilization are inducible. Their synthesis by the structural genes does not occur in the absence of an inducer because the regulatory gene i^+ produces a repressor substance that prevents synthesis. However, when an inducer substance is present, the repressor substance is inactive and enzyme synthesis occurs. Finally, the o^+ gene has the property of responding to the repressor substance produced by the i^+ gene.

ENZYME REPRESSION In addition to inducible systems, such as the system for lactose utilization, there are repressible systems. Enzyme synthesis can be repressed in the presence of a *corepressor*, which is either some exogenously supplied substance or some substance produced within the cell. This substance can be the product of the reaction or reactions catalyzed by the enzymes that

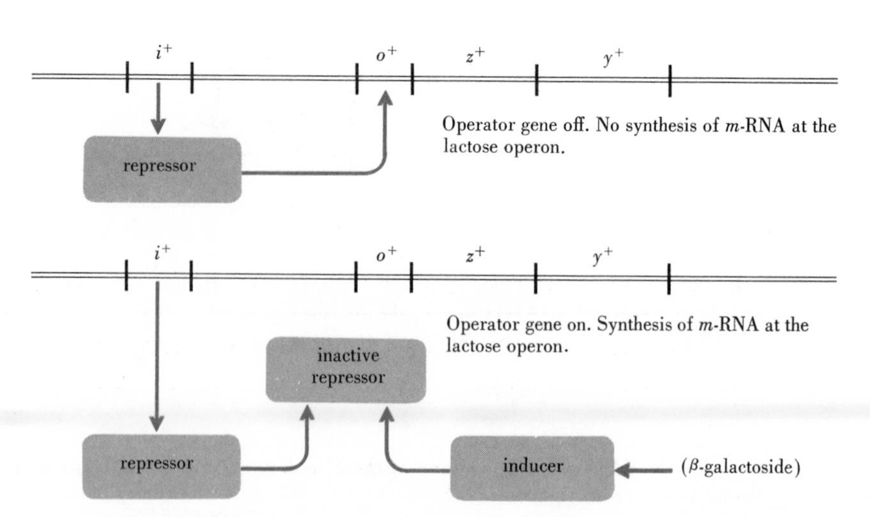

Fig. 11-2 Regulation of the lactose operon (see text for details).

are synthesized by structural genes. In a repressible system, the product of a regulatory gene, in combination with the corepressor, has the effect of sending a signal that is recognized by an operator gene, and this recognition turns off the activity of a structural gene. An example of a repressible system is one involving histidine biosynthesis in *E. coli*. At the appropriate concentration of histidine the synthesis of all of the enzymes required for histidine biosynthesis is repressed. Thus, histidine is acting as the corepressor. This is called *coordinate repression*. Provided with histidine the cells cease synthesizing enzymes that are now unessential. Just as in the case of an inducible system, such as for lactose utilization, the cells do not synthesize enzymes whose function is not immediately required. Figure 11-3 compares a repressible system with an inducible system. The only difference between the two systems is that in an inducible system the repressor by itself combines with the operator to turn off synthesis at a structural gene, whereas in a repressible system the repressor is inactive unless combined with a corepressor. In this form the repressor combines with the operator to turn off synthesis at a structural gene.

FEEDBACK INHIBITION Enzyme synthesis can be subjected to genetic regulation, and this is one way in which a cell's metabolism can be affected. There is, however, another way to

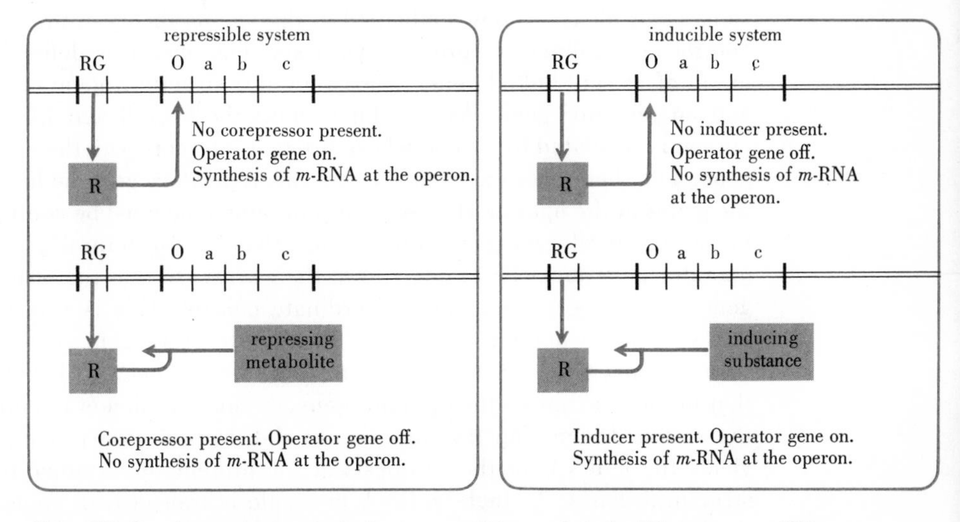

Fig. 11-3 *A comparison of the repressible and inducible systems.* RG = *regulatory gene;* O = *operator gene;* a, b, *and* c = *structural genes;* R = *product of* RG.

affect metabolism at the enzyme level—that is, by the *inhibition* of the activity of enzymes already present in the cell. This inhibition is called *feedback inhibition,* for the end product of a biosynthetic pathway inhibits the activity of the first enzyme of the pathway. For example, the first step in the biosynthesis of histidine is catalyzed by the enzyme pyrophosphorylase, and in the presence of the appropriate endogenous concentration of histidine the activity of the enzyme is inhibited. When the histidine concentration decreases to a certain level, the enzyme becomes active again. The histidine binds to a site in the protein and the conformation of the protein changes. The change results in a decrease in enzyme activity. The active conformation is regained after the bound histidine leaves the enzyme.

Feedback inhibition is related to induction and repression in that it is an *on-off* phenomenon, but it is a nongenetic way in which a cell can regulate its metabolism at a rate appropriate to both external and internal environmental conditions.

THE SIGNIFICANCE OF GENETIC REGULATION

At this point you should have begun to consider the role that regulatory systems can play in the life of a cell and how these systems might lead genetically identical cells down different developmental pathways. In 1960 Jacob and Monod and their co-workers began to generalize some of their ideas regarding enzyme induction and repression. Their ideas are now embodied in the *operon* theory of the genetic control of regulation of protein synthesis. The operon is defined as a group of closely linked genes consisting of a structural gene or genes and an operator gene. As you know now, the overall activity of the operon is regulated by a gene whose product interacts with the operator gene. The theory does not require that this regulatory gene be linked to the genes of the operon. However, the operator gene must be contiguous to its structural genes; if there is more than one structural gene, the operator gene must be at one end of their array, so that the structural genes of an operon are under coordinate control. This is true of the genes of the *lac* operon and also for the nine structural genes concerned with the enzymes of histidine biosynthesis in *Salmonella.* It is believed that in most instances the operator gene, or some portion of it, contains the codon designating the initiation of the transcription process that yields the *m*-RNA for the synthesis of the proteins determined by the structural genes. A single *m*-RNA molecule is transcribed, containing the information for all of the proteins determined by all of the structural genes, rather than a separate *m*-RNA molecule for each of the proteins. In the case of the histidine operon of *Salmonella*, the operator gene maps at one end of the operon and regulates the synthesis of an *m*-RNA that results from the transcription of all ten structural genes. The same is true for the operator gene in the lactose operon.

To date most of our knowledge of the genetic regulation of enzyme synthesis stems from investigations with bacteria, and control networks as diagrammed in Fig. 11-3, have their origin almost exclusively from the investigation of enzyme systems in bacteria. Do genetic regulatory systems exist in other organisms, specifically among nucleate or higher organisms? Examples of induced and repressed enzyme synthesis are known for a variety of different nucleate organisms, and it might be assumed that regulatory genes are present in these organisms to govern enzyme synthesis. However, there is, so far, no clear evidence for the existence of genes with regulatory properties. On the other hand, there is evidence for clusters of genes concerned with biosynthesis of certain amino acids, notably, structural genes for the biosynthesis of aromatic amino acids in *Neurospora* and for histidine biosynthesis in *Aspergillus*, *Neurospora*, and *Saccharomyces*. Presumably, these clusters of structural genes constitute portions of operons, but the precise description of these operons depends upon identifying their regulatory genes.

Clearly, genetic regulation can play an important role in differentiation at the level of a single-celled organism, and undoubtedly it is of equal importance in development and differentiation among higher organisms. The synthesis of enzymes is under genetic control in higher organisms, but the nature of the control systems is just beginning to be understood.

FURTHER READING

McFall, E., Maas, W., 1967. "Regulation of enzyme synthesis in microorganisms," in *Molecular genetics*, part 2, p. 255, Taylor, J. H. (ed.). New York: Academic Press Inc.

INDEX

A

E

Y

Z